Water and Soil in Holy Matrimony?

A smallholder farmer's innovative agricultural practices for adapting to climate in rural Zimbabwe

Christopher Munyaradzi Mabeza

Langaa

Langaa Research & Publishing CIG
Mankon, Bamenda

Publisher
Langaa RPCIG
Langaa Research & Publishing Common Initiative Group
P.O. Box 902 Mankon
Bamenda
North West Region
Cameroon
Langaagrp@gmail.com
www.langaa-rpcig.net

Distributed in and outside N. America by African Books Collective
orders@africanbookscollective.com
www.africanbookscollective.com

ISBN-10: 9956-764-51-5

ISBN-13: 978-9956-764-51-8

© Christopher Munyaradzi Mabeza 2017

Declaration

This book is a product of my PhD thesis. Where contributions of others are involved, every effort is made to indicate this clearly, with due reference to the literature and acknowledgement of collaborative research and discussions.

Dedication

This book is dedicated to the memory of
Andrew Sifelani Mabeza –
my young brother and pillar of strength –
who sadly passed on in 2007.
May his soul rest in eternal peace.

Table of Contents

List of Illustrations

List of Tables

List of photographs

Chapter 4

List of acronyms

AGRA	Alliance for a Green Revolution in Africa
AGRITEX	Agricultural Technical and Extension Service
AIDS	Acquired Immune Deficiency Syndrome
BSAC	British South Africa Company
CA	Conservation Agriculture
CARE	Cooperative for American Relief Everywhere
CNN	Cable News Network
CSA	Climate Smart Agriculture
CSO	Central Statistical Office
DA	District Administrator
Ed	Editor
Eds	Editors
ENDA ZIMBABWE	Environmental Development Activities - Zimbabwe
EPOPA	Export Programme for Organic Products from Africa
FAO	Food and Agriculture Organization
GHG	Greenhouse Gases
HIV	Human Immunodeficiency Virus
HQ	Headquarters
IEG	Independent Evaluation Group
IPCC	Intergovernmental Panel on Climate Change
IPCC TAR	Intergovernmental Panel on Climate Change Third Assessment Report
LAA	Land Apportionment Act
MACO	Midlands AIDS Caring Organisation
MDC	Movement for Democratic Change
MVP	Millennium Villages Project
NGO	Non-Governmental Organisation
NGWP	New Gato Water Project
NLHA	Native Land Husbandry Act

PELUM	Participatory Ecological Land Use Management
PROLINNOVA	Promoting Local Innovation in Ecologically-Oriented Agriculture and Natural Resource Management
POSA	Public Order and Security Act
RDC	Rural District Council
RIU	Research into Use
R&D	Research and Development
SADC-REEP	Southern African Development Community Regional Environmental Education Programme
SREX	Special Report on Managing the Risks of Extreme Events and Disasters
TA	Transformational Adaptation
TLA	Traditional Leaders Act
TOT	Transfer of Technology
UNDP	United Nations Development Programme
UNFCCC	United Nations Framework Convention on Climate Change
UK	United Kingdom
USAID	United States Agency for International Development
USD	United States Dollar
VIDCO	Village Development Committee
WDP	World Development Programme
WADCO	Ward Development Committee
ZACT	Zimbabwe AIDS Caring Trust
ZANU PF	Zimbabwe African National Union – Patriotic Front
ZEPPDRA	Zimbabwe Ex-Political Prisoners Detainees and Restrictees Association
ZMSD	Zimbabwe Meteorological Services Department
ZWD	Zimbabwe Dollar
ZWP	Zvishavane Water Project

Acknowledgements

This book is a product of my PhD thesis based on an ethnographic fieldwork in rural Zimbabwe.

I am very deeply indebted to my supervisors A/Prof Lesley Green, A/Prof Gina Ziervogel and A/Prof Shadreck Chirikure for the guidance they gave me through the entire journey. Thanks for their patience and commitment that enabled me to write this story of Mr Phiri Maseko's 'marriage of water and soil'. Their contributions to this book are greatly appreciated. Many thanks to Prof Francis Nyamnjoh and Dr Frank Matose both from the University of Cape Town for their wonderful insights that have also helped shape this study. Frank has been an inspiration since we first met in high school. I also want to acknowledge the following scholars from the University of Zimbabwe: Dr M Musiyiwa, Prof G Kahari and Mr A Daimon for the insightful discussions we held. I also wish to convey my appreciation to Prof R Auerbach of Nelson Mandela Metropolitan University and Dr C van Rooyen of the University of Johannesburg for their constructive comments.

My most heartfelt thanks go to the Andrew Mellon Foundation for funding my studies. I was awarded the Sawyer Seminar Doctoral Fellowship in 2011 and the AW Mellon Doctoral Fellowship from 2012 to 2014. I am grateful to A/Prof Lesley Green for facilitating these fellowships. Without the generous awards by the AW Mellon Foundation, this study would not have been possible.

I wish to convey my deep gratitude to Mr and Mrs Phiri Maseko for allowing me to stay with them during my ethnographic fieldwork. I recall with fond memories the wonderful moments we shared together. Sadly, on the 1st of September 2015, Mr Phiri Maseko passed on. This was long after this study had been completed. May his soul rest in eternal peace. I wrote a tribute for the late Mr Phiri Maseko (see Appendix 1).

Special thanks also go to Dr Ken Wilson and Prof Ian Scoones for their insightful research work with Mr Phiri Maseko in rural Zvishavane. I want to thank Ken for compiling Mr Phiri Maseko's Book of life and affording me an opportunity to visit Mr Phiri

Maseko's adopters in Mazvihwa. Ken also shared with me his insightful thoughts about Mr Phiri Maseko's agricultural practices that I use in this book. I am also indebted to all the research partners I interacted with in Zvishavane during my fieldwork.

It was a pleasure working with the following colleagues in the PhD Contested Ecologies team: Josh Cohen, Kelsey Draper, Munya Mawere, Art Nhemachena and Marieke Norton. I found our discussions to be very scintillating. I greatly appreciate the discussions I held with Witness Kozanayi that yielded fascinating insights. Over the years, I have been honoured to have friends and relatives who have rendered me help in different capacities. To that end, I want to thank the following: Raphael Mashingaidze, Billy Mukamuri, Kudzai Sikweche, Steve Matema, Ralph Chirowamhangu, Washington Zhakata, Glen Ncube, Lawrence Chanaka, Arnold Sithole, Torevei Kurasha, Bernard Madziwa, Bessie Madziwa and my brother Anthony and sister-in-law Pricilla Mabeza.

Staying in Cape Town was like being home away from home, thanks to the following friends who gave me wonderful company: Saviour Adgenti, Xolisa and Bandi Mnyanda, Duncan and Susan Mhakure, Andy Makhalima, Ordain Hove, Brian Hove, Frank and Tammie Matose and James and Rose Irlam.

To my parents, Mr Andrew and Mrs Miriam Mabeza, I am deeply indebted for your love, support and encouragement. To my siblings Caro, Simba, Tsitsi and Sibby, my sister-in-law Mai Pana (Pana's mother) and my son Chris Jr, many thanks for encouraging me to reach for my stars. Lamentably, my sister-in-law Mai Vee (Vee's mother) could not live to witness this project come to fruition. May her soul rest in eternal peace. To my nephew Panashe and nieces Cher, Gamu, Vee, Natasha, Vongai and Miracle, thank you for your prayers.

Finally, yet importantly, I want to thank my God for His divine guidance and protection. Without God's love, I would not have completed this journey - a journey that had all the hallmarks of relentless tenacity.

Out of the huts of history's shame
I rise
Up from a past that's rooted in pain
I rise
I'm a black ocean, leaping and wide,
Welling and swelling I bear in the tide.
Leaving behind nights of terror and fear
I rise
Into a daybreak that's wondrously clear
I rise
Bringing the gifts that my ancestors gave,
I am the dream and the hope…
I rise
I rise
I rise

-    Maya Angelou (1928-2014)

137

Foreword

Rainwater harvesting for sustainable rural agriculture in tropical Africa is a rare feat, where this has happened. Dr Chris Mabeza's book examines the success story of Mr Zephaniah Phiri Maseko who engaged in a unique rainwater harvesting system in which water from the small high area and hill atop his homestead is meticulously channelled down his land below, systematically watering land naturally via percolation at the same time without causing soil erosion. The water slowly seeps into the undulating slopes, wetting the whole field and at the same time creating a zone of intermittent saturation not far down from the visible top soils. Consequently, Mr Phiri Maseko's land often has sufficient moisture supply for agricultural practices even before the formal rain season starts. He is then able to plant his crops earlier than others, and in the dry season, he can still harvest because of the sophisticated underground water harvesting system that he has created over the years.

As a person who was born in the district of Zvishavane, which is generally an agro-region 4 area, (which receives less than 600mm of rain per year), I had the opportunity to see for myself this project at work way back in the early 1990s. Mr Phiri Maseko's smallholder farm is in the ordinary Runde Rural communal areas. His physical environment is not more unique to that of his neighbours or other rural Zimbabweans. What is unique is the level to which he has applied his ingenuity to African indigenous knowledge systems and at the same time, his sophisticated knowledge of modern agriculture, soil conservation and rainwater harvesting. His invention has earned him so many accolades, including international recognition for this work. He has partnered with some development agencies and funders to start what has now become the Zvishavane Water Project. The project aims to replicate Mr Phiri Maseko's model in the district and beyond. Several international researchers have shown interest in Mr Phiri Maseko's work, but no one has yet published a book on this innovative work. Dr Mabeza's work is therefore timely and important.

Scrupulously researched, this ethnographic study is not only a study about the person who has done the invention, but also crucially, it is about how rural Africans could rethink and adapt to climatic changes, particularly in areas of increasing aridity. Such adaptation and innovation does not necessarily always require western intervention, but require local African ingenuity and solution finding.

Dr Mabeza's work is a cutting edge anthropological examination of African rural socio-ecology. It is deeply engaged in its understanding of contemporary literature on environmental anthropology and rural sociology, with critical discussions on issues of resilience, climatic change, and smallholder agriculture strongly foregrounded. As Chris Mabeza argues, "innovations by smallholders play a complementary role to interventions by outsiders in the discourse of adaptation to climate in the drylands of southern Zimbabwe." In the face of increasing aridity, Mr Phiri Maseko has managed to "marry water to the soil" in such a way that the water does not easily 'elope' downstream, but keeps an enduring contact with the soil to ensure that his family does not lack food. This interdisciplinary work is one of the major works on environmental anthropology and rural sociology in Zimbabwe. I strongly recommend it.

Prof Enocent Msindo (PhD, Cambridge)
Deputy Dean of Humanities,
Rhodes University, South Africa.

Introduction

> They had learned nothing and forgotten nothing – Talleyrand in speaking about the restored French Bourbon dynasty after the abdication of Napoleon Bonaparte - St Augustine's Press, 2016

"Nothing learned, nothing forgotten"

Old habits die hard, or is it that old dogs are hardwired not to drool at the prospect of new tricks? Or both. Whatever the case might be, interventions in rural areas have been a source of fierce debates. Nowhere is this evident than in the rural development discourse in Zimbabwe where smallholder farmers face an uncertain future as the vagaries of a changing climatic environment take root. Smallholders grapple with food insecurity. Food security has been defined as, "the success of local livelihoods to guarantee access to sufficient food at household level (Devereaux and Maxwell cited in Ziervogel et al., 2006). Conversely (and taking a cue from the above definition), food insecurity is the inability of local livelihoods to make sufficient food accessible to households. The rural landscape is replete with interventions meant to help rural communities become food secure. Despite these so-called interventions of good will, smallholders remain trapped in a vicious poverty cycle. Locked inside their echo chamber modus operandi, players in the rural development discourse seem not to get their act together. Like the Bourbon dynasty in the days of yore, the proponents of the development agenda appear to have learned nothing and forgotten nothing about the errors of their ways. Resultantly, this has led to an academic stampede as scholars wrestle with the underlying causes of perennial rural food insecurity.

For example, in a paper presented at the *Farmer First Revisited: 20 Years On* conference at the Institute of Development Studies, University of Sussex, UK in December 2007, one of the key speakers, Andy Hall asks the question: "Why are we still here?" (Hall, 2009: 30). He uses this question as "lens to look at the challenges to strengthening agricultural innovation" (Hall, 2009: 30). About twenty years before this conference, Andy Hall had attended the *Farmer First Conference* as a PhD candidate working on agricultural innovations (see also Scoones and Thompson, 2009). Young and naïve as he was then, he thought they "had the problem sorted out". Yet twenty years later, the same challenges have persisted. Andy Hall's question reverberates throughout this book. Smallholder farmers continue to face monumental challenges in a bid to adapt to climate variability despite being overwhelmed over the years by interventions by both state and non-state actors. Why are they perpetually embedded in a state of food insecurity? Why are they still 'there'? Why have smallholder farmers remained vulnerable to the adverse effects of climate variability?

Study area

This study echoes Andy Hall's question in an attempt to address smallholders' barriers to climate adaptation. The research highlights the inspirational work of Mr Zephaniah Phiri Maseko, an internationally acclaimed and award-winning smallholder farmer who initiated innovative agricultural practices as a response to climate variability in drought-prone Zvishavane, rural Zimbabwe.

The study area, Zvishavane is in semi-arid southern Zimbabwe. It comprises nineteen wards and the study site is in Ward 6, known as Mapirimira (see Map 3). This book will use both terms Mapirimira Ward and Ward 6 interchangeably. The ward is about 18 km to the north west of Zvishavane town and falls under the jurisdiction of Chief Mapanzure of the *Va*Mhari people. The total population of the ward is 4352, comprising 1980 males and 2372 females (Zimbabwe National Statistics Agency, 2012). The total number of households is 966 with an average of 4.5 people per household (Zimbabwe National Statistics Agency, 2012).

Zvishavane is in agro-ecological zones 4 and 5 based on agricultural potential (see Map 1) (Vincent and Thomas, 1960). Agro-ecological region 4 is characterised by semi-extensive farming. Rainfall is low and variable with intra-seasonal dry spells. In agro-ecological region 5, rainfall is extremely variable and not even reliable for drought resistant crops. Rainfall average in Zvishavane is below 600mm (see Figure 1).

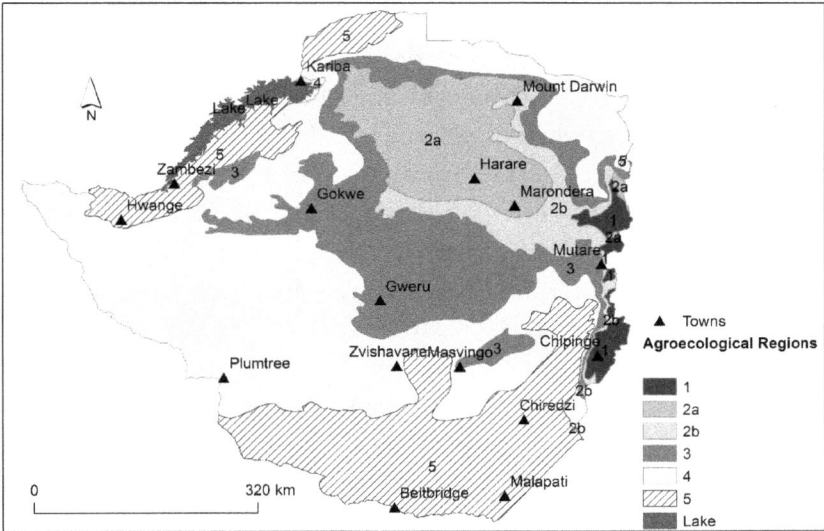

Map 1: Agro-ecological zones of Zimbabwe (Source: Murwira et al., forthcoming)

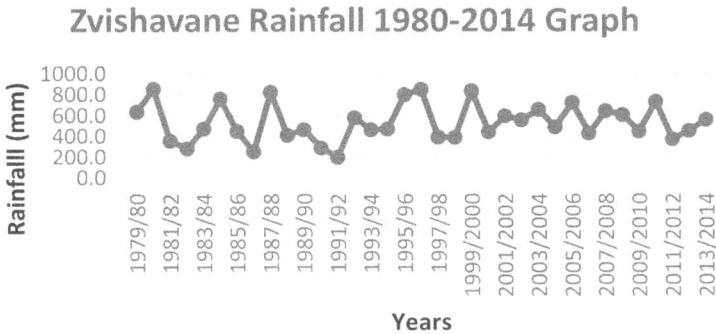

Figure 1: Zvishavane rainfall graph from 1980 to 2014 (Source: Zimbabwe Metrorological Services Department {ZMSD}, 2014)

Smallholder farming is the major economic activity in the ward, however, some villagers engage in off-farm livelihood strategies such as basket weaving and artisanal mining of gold. The villagers mainly engage in mixed farming with most growing maize, groundnuts, sorghum and millet. Some engage in market gardening and grow crops such as tomatoes, cabbages and onions. Three extension officers from the Agricultural Technical and Extension Service (AGRITEX) work in Mapirimira Ward. AGRITEX divided the ward into three clusters. A cluster is made up of seven villages. Each of the extension workers covers one cluster. Mr Phiri Maseko lived in Ziyabangwa village, which is adjacent to Hlupo village. Both villages are mainly inhabited by the *Va*Mhari of the *shumba* (lion) totem. Ziyabangwa is an off-shoot of Hlupo village which had grown too large for administrative purposes. Mr Sibanda (village head of Ziyabangwa) says that Chief Mapunzure's area used to be inhabited by the Rozvi people. The *Va*Mhari are said to have come from neighbouring Chivi District about two hundred years ago, and they fought and defeated the inhabitants of the area, the Rozvi, before settling in present day Mapanzure area. The chief and village heads run the day-to-day affairs of the area. However, with the advent of the colonial land policy of the 1930s, the people were moved from traditional areas they had inhabited into reserves that are now referred to as Zvishavane rural or Zvishavane Communal Lands where Mr Phiri Maseko lived.

Land question

Zimbabwe's land question is one of the most contentious issues in the country's history. However, I suspect brickbats might be thrown in my direction for not discussing the controversial fast track land policy. It is my submission that the main research subject is a farmer in a communal area and therefore, the fast track land resettlement programme is not within the purview of this study. Thus, the study puts the history of the land question into context.

Reserves were created to make land available for white commercial farms. The context in which rural communities respond to interplay of stressors including changing climate cannot be fully understood if consideration is not given to the history of Zimbabwe's land question. The British colonised Zimbabwe in 1890 resulting in extreme change in land ownership structures among the new settlers and black Zimbabweans. European farming in Rhodesia (as Zimbabwe was known during colonial times) inherited features from the South African model mainly because most of the settlers were born in South Africa or had spent time there (Palmer, 1977). European settlers could 'ride off' in typical South African style; in other words, demarcation of farms was done by riding horses for long distances and then peg the boundaries of the farm much to the detriment of the African people who would be displaced in the process. This type of demarcation was erroneously premised on European farms in the Cape which were vast because the land was more suitable for ranching and characterised by general lack of water (Palmer, 1977). The settlers in Rhodesia also followed the South African example in creating reserves. In South Africa, reserves had been created to accommodate dispossessed 'Hottentots' in the Cape, and the same principle was applied in Natal and Transkei. It must be noted that reserves were not designed by the colonial powers to be economically viable. Rather, they were created to be labour reservoirs to serve colonial economic interests in the new urban industrial areas and commercial farms (Van Onselen, 1976; Phimister, 1988). Fundamentally, the settler land policy "guaranteed white economic dominance and black poverty during the colonial period" (Herbst, cited in Rukuni et al., 2006). Woddis concurs and posits that land expropriation was meant:

> to prevent the African peasant from becoming a competitor to the European farmer or plantation owner; and to impoverish the African peasantry to such an extent that the majority of adult males would be compelled to work for the Europeans, in the mines or on the farms (Woddis, 1960: 8).

The British colonial settlers having formed the British South Africa Company (BSAC) in 1889 crossed the Limpopo into present day Zimbabwe from South Africa in pursuit of a 'Second Rand'. Little did they know their assertions were incorrect, as there were no large-scale gold deposits in the country. The lack of gold resulted in

BSAC shares plummeting on the London Stock Exchange. To restore the value of the shares, the white settlers turned to agriculture and began to expropriate land from black Zimbabweans (Phimister, 1988; Rukuni et al., 2006). After the 1893-94 Anglo-Ndebele war, the Land Commission of 1894 created two reserves, Gwaai and Shangani, for the vanquished Ndebele, and thus land alienation was set in motion. In 1908, the settler government embarked on a white agricultural policy. *Chibharo* (forced labour) was introduced and many African people were forced to abandon their traditional ways of survival and forced to work for the settlers as farm and mine labourers. This necessarily undermined African agricultural systems and the control of land by the white settlers became the vehicle by which Africans were controlled economically and politically (Palmer,1977).

Palmer (1977) contends that the issue of land alienation being driven by the need for white dominance over blacks become apparent in later years but cannot be used to explain the expropriation of land in the 1890s. Palmer further argues that:

> The BSAC had been founded in 1889, at Rhodes's insistence, expressly for the purpose of exploiting the lands to the north of the Limpopo and so compensating Rhodes and his associates for their failure to strike it rich on the Rand. Rhodes also anticipated that the rich Mashonaland would act as a counter weight to the Transvaal, and so reassert British political supremacy in South Africa. The company's attentions, therefore, were focused firmly upon the rapid discovery and exploitation of mineral resources; all else, and especially 'native policy', was of secondary consideration (Palmer, 1977: 25).

The land question is one of Zimbabwe's most enduring and highly contested legacies (Kinsey, 2004). The importance of land in Shona cosmologies can never be doubted. Therefore, there was always potentially conflict when colonial legislation appropriating land from black Zimbabweans was passed.

The Land Apportionment Act (LAA) of 1930 resulted in the British creating large commercial farms for settlers by taking away land from the majority black population. Most of the land that was expropriated for white farmers was fertile and located in areas with high rainfall. Africans were forced to settle in overcrowded and agriculturally marginal 'reserves'. With the increase in population, land alienation as well as few employment opportunities led to overcrowding in the rural areas and this resulted in land degradation (Elwell, 1985; Whitlow, 1988; Lliefe, 1990; Mehretu and Mutambirwa, 2006). Lliefe (1990) argues that from 1930, land alienation and an increase in population reduced grain production by Africans and further suggests that scarcities were mainly felt in southwestern Zimbabwe that had been greatly disrupted by European settlement. It is against the background of such an environment that smallholders such as Mr Phiri Maseko rose to prominence through innovations that would mould him into the traditional Shona *hurudza*, regardless of the infertility of the soil in areas allocated to Africans.

A total of 21 127 040 acres were allocated to Africans and 49 149 174 for Europeans (see Table 1). In 1930 the African population was about 1 081 000 and that of Europeans was close to 50 000 but 51% of the land was assigned to the whites and 29.8% to Africans (Moyana, 1984). Scholars such as Gann (1934) have hailed LAA as a legal statute that did not alienate land from the African people. Gann cited in Moyana argues:

> The Act provided the Africans with an area only slightly smaller than that of the whole of England. The indigenous folk in terms of land thus fared a great deal better than the Red Indians of North America, the Maori of New Zealand or the Araucanlans of Chile (Gann cited in Moyana 1984: 280).

However, Gann's analysis ignores the fact that the land allocated to Africans had impoverished soils and in semi-arid areas. This continued to be a festering wound in relations between the races.

Category	Acres	% of country
European Area	49 149 174	51.0
Native Reserves	21 127 040	22.0
Unassigned Area	17 793 300	18.5
Native Area	7 464 566	7.8
Forest Area	590 500	0.6
Undetermined Area	88 540	0.1
Total	96 213 120	100.0
Total for African use	28 591 606	29.8

Table 1: Land Apportionment in Southern Rhodesia (Zimbabwe) in 1930 (Source: Moyana, 1984:70)

Zimbabwe's land tenure (see also Table 2) is mainly communal albeit this in some respects can be inaccurate because this implies "common ownership of all resources and collective production, which is hardly found" (Cousins, undated: 152). Cousins elaborates:

> ...communal means, in the great majority of cases, a degree of community control over who is allowed into the group, thereby qualifying for an allocation of land for residence and cropping as well as rights of access to and use of the shared common pool resources used by the group (i.e. the commons). Groups often restrict alienation of land to outsiders and thus seek to maintain the identity, coherence

and livelihood security of the group and its members. However, allocations of residential and arable land usually result in strong rights for individuals or families or both who sometimes also exercise rights over land which contains common pool resources such as water points, or wetter areas with dry season grazing. These can also be controlled by sub-groups (kinship groupings, or clans) within larger groups (Cousins, undated).

Access ('rights')	Control ('authority')
Derived from membership of nested social units (e.g. household, kinship groups, community)	Guarantees access
Acquired via birth affiliation, transactions	Regulates common property use
Relative	Redistributes access
Shared	Resolves disputes
Inclusive	Nested within levels of socio-political authority (e.g. 'tribe', 'ward', 'village')

Table 2: Key features of African land tenure (Adapted from Cousins, undated)

In 1951 the colonial authorities in(then) Rhodesia enacted the Native Land Husbandry Act (NLHA). NLHA was based on the principle of private individual ownership of land and government sponsored soil conservation techniques (Moyana, 1984; Mlambo, 2009: 106). These techniques were supposedly meant to halt soil erosion and land degradation in the communal areas that had been caused by the LAA of 1931. Before the creation of reserves, the farming communities practised transhumance, meaning that they could graze their cattle in a large area according to the availability of pastures allowing for regeneration. With the promulgation of the NLHA, they could only graze their cattle in the same area leading to overgrazing and soil erosion.

The provisions of the NLHA included compulsory contour ridging and de-stocking (Mlambo, 2009: 106). The thinking was that de-stocking and contour ridging would halt land degradation. The contour ridges might have exacerbated the problem of land degradation. Mr Phiri Maseko realised that these contour ridges diverted water out of the fields and by so doing led to the formation of gullies. The NLHA was "a prescription for a wrong illness" (Moyana, 1984: 136). Mr Phiri Maseko also emerged out of this set-up of colonial land management and tenure system. Water management is central to his innovative agricultural system. He re-invented the colonial contour ridges. His deepened contour ridges store water rather than drain the water away. It is against these historical precedents that smallholders are grappling with issues to do with climate variability in semi-arid southern Zimbabwe and other parts of Africa. Compulsory construction of anti-erosion structures during the colonial era was not only confined to Zimbabwe and can be found elsewhere in Africa.

During the colonial era in Africa, scientific framing of erosion was always prominent in state interventions (Weisser et al., 2014). For example, scientific framing of soil erosion in Rwanda led to forced resettlement and was premised on construction of anti-erosion structures through forced labour by Belgian authorities (Weisser, 2014). In Zimbabwe, compulsory contour ridging was fiercely resisted by the African people in colonial Zimbabwe. Opposition to the NLHA is captured in song. One of the songs composed by Zimbabwean music icon, Thomas Mapfumo is entitled *Makandiwa* (Contour Ridges) and in the song, he laments the suffering that ensued during the forced construction of the contour ridges (with the English translation):

Makandiwa

Nhamo yamakandiwa takaiona he nhayi Mambo
Nhamo yamakandiwa takaiona he nhayi Mambo
Nhamo yamakandiwa takaiona he nhayi Mambo
Nhamo yamakandiwa takaiona he nhayi Mambo

Contours

We suffered constructing contour ridges dear Lord
We suffered constructing contour ridges dear Lord
We suffered constructing contour ridges dear Lord
We suffered constructing contour ridges dear Lord

Construction of contour ridges was vehemently resisted by black Zimbabweans and even during my fieldwork Mr Phiri Maseko retained sad memories about this era. He said it was out of the realisation that the contour ridges drained water out of his plot that led to his innovation of deepening the contours for them to store water.

There was also another song during the 1950s that critiqued NLHA entitled *Mombe mbiri nemadhongi mashanu,* (Two oxen and five donkeys):

Mombe mbiri nemadhongi mashanu

Mombe mbiri namadhongi mashanu sevenza nhamo ichauya
Mombe, mombe mbiri namadhongi mashanu sevenza nhamo ichauya
Sevenza, sevenza nhamo ichauya
Sevenza, sevenza nhamo ichauya
Mombe mbiri namadhongi mashanu sevenza nhamo ichauya
Mombe, mombe mbiri namadhongi mashanu sevenza nhamo ichauya
Sevenza, sevenza nhamo ichauya
Sevenza nhamo ichauya

Two cows and five donkeys

Two cows and five donkeys, one must work hard to overcome poverty caused by destocking
Two cows and five donkeys, one must work hard to overcome poverty caused by destocking
Work hard to overcome poverty
Work hard to overcome poverty
Two cows and five donkeys, one must work hard to overcome poverty caused by destocking

Two cows and five donkeys, one must work hard to overcome poverty caused by destocking
Work hard to overcome poverty
Work hard to overcome poverty

The song *Mombe mbiri namadhongi mashanu* is a social commentary about compulsory destocking, one of the provisions of the NLHA. The song implores people to work hard with the few resources available if they are to avoid famine. This compulsory destocking wreaked havoc among the rural population. Mr Phiri Maseko recalls that before the NLHA came into effect; his father had close to 30 cattle and ten donkeys. These stocks were reduced to about four cattle and five donkeys by 1953. He recalls that his father was forced to slaughter many of his cattle and sold the meat very cheaply so as to reduce the number of livestock to the stipulated figure as per the provisions of the NLHA. These provisions of the NLHA greatly angered Mr Phiri Maseko and many other rural people and this might have influenced him to join the liberation struggle. He said he was angry when his father sold livestock at low prices in trying to comply with the NLHA. Mr Phiri Maseko decided to join the National Democratic Party (one of the first political parties to be launched by black Zimbabweans) when it was launched in the late 1950s with the hope that colonial legislation like the NLHA would be annulled in an independent Zimbabwe.

Mr Phiri Maseko emerged out of this grossly inequitable land management tenure structure during the colonial era as already alluded to. The colonial contour ridges promoted siltation because soil was washed away by water resulting in the now familiar sight of dry river beds, a situation also observed by colonial administrators at the time. The Report of the Secretary for Internal Affairs for the Year 1963 reads:

> Owing to lack of funds, no measures were taken for conserving or increasing water supplies. This is a serious position, for without conservation measures this district will, in time, become uninhabitable owing to lack of water. Underground supplies cannot be expected to

last forever without some conservation measures being taken. Money must be founded somewhere and in the not-too distant future (Secretary for Internal Affairs, 1963: 10).

The Southern Rhodesian "Report of the Secretary for internal affairs for the Year 1963" says:

> In some areas, district commissioners attribute the lack of progress to political unrest that has resulted in extension advice being spurned, contour ridges ignored or training centres closed for lack of interest (Secretary for Internal Affairs, 1963: 10).

This illustrates the "everyday forms of resistance" (as Scott, 1985 would say) to colonial land management policy as evidenced by the African people simply refusing to dig the contour ridges. However, Mr Phiri Maseko's resistance to colonial rule manifested itself in the form of challenging colonial environmental policy to enhance his livelihoods. He recognised that agro-ecological region 4 requires what he calls deepened contours, which preserve water. He dug infiltration pits in the contour ridges that are now known as the 'Phiri Pits'. Along the contour ridges he planted both fruit and non-fruit trees. The trees hold the soil from falling into the contour ridges. Mr Phiri Maseko said that if one plants water in the soil, it resists run-off, helping farmers realise high yields, a stark reminder of the role the soil plays in Shona worldviews. But what is the significance of the soil to the Shona of Zimbabwe?

Shona cosmology of water, soil and marriage

Mr Phiri Maseko's notion of marriage of water and soil resonates with southern African thought more broadly as grounded on a philosophy of soil and relationalities. For instance, his practices reflect both Malawian and the Shona of Zimbabwe's conceptualisation of water and soil. In this study, I use the words land and soil interchangeably as it is my assertion that there is a very thin line separating the two. Other scholars who have written about the

Shona land philosophy who include Moyana (1984), Bourdillon (1987) and Musiyiwa (forthcoming) share this view. The relationship between the Shona and the land is anchored in many beliefs. The synergy as Bourdillon (1987: 67) observes, has however "radically changed". The economic crisis in Zimbabwe has led to the closure of many industries pushing the unemployment rate up. This has increased pressure on land especially in the rural areas. Bourdillon (1987: 67) maintains that in such a scenario the "land's economic value and the significance of rights of individuals of land become prominent, even the old ideas about land are not forgotten".

The Shona believe that the land is the realm of the ancestors (Bourdillon, 1987; Musiyiwa, forthcoming). Musiyiwa says that in order to understand the philosophical underpinnings of the Shona interaction with the soil, it is imperative to acknowledge that religion is at the core of the Shona culture. It is the central institution from which other institutions justify their existence. Religion is a powerful establishment that regulates everything, including how people interact with the land and water. This interaction entails a relationship linked by an unbroken umbilical cord. Musiyiwa adds:

> ...the Shona people's natural relationship with the land, the relationship enshrined in a historically celebrated land mythology...Although the land mythology is a component of a broader Shona mythology, it is its most integral component which amounts to an indigenous land philosophy. It encompasses virtually the entirety of Shona ontology which revolved around land (Musiyiwa, forthcoming).

The Shona regard the soil as sacred. Given this significance, there exist many taboos connected to the soil. For instance, during the *chisi,* a sacred day (usually a Thursday or Wednesday) observed every week, the Shona are not supposed to work on the land as a sign of paying homage to the ancestors who inhabit it. If anyone is found flouting the rules governing the observance of the *chisi,* sanctions are applied by the village head in consultation with the chief. The Shona also believe that failure to observe the *chisi* may result in drought or pandemics. Mawere and Wilson (1995) examine the impacts of a socio-religious movement in the Mazvihwa area of rural Zvishavane in 1992 led by the Ambuya Juliana movement. At the heart of the Ambuya Juliana movement was the declaration of a new taboo regime in the Mazvihwa area aimed at "re-establishing socio-ecological order" (Mawere and Wilson, 1995: 257). One of the taboos stipulated that people in the area were to observe Wednesday as the *chisi*. Sanctions against those who violated the taboos included fines in the form of livestock. The Ambuya Juliana movement had a sizeable following in the Mazvihwa area further demonstrating the centrality of ancestors in the lives of the Shona.

Founders of the chieftaincies in Shona cosmologies are the *vene vevhu* (owners of the soil) since their remains lay buried in the soil of their specific territories. Bourdillon points out that,

> 'Ownership' of the land by the spirits is bound up with the relationship between the spirits and the living community. The land forms a close and enduring bond between the living and the dead: through their control of the fertility of the land they once cultivated, the spirits are believed to continue to care for their descendants and the descendants are forced to remember and honour the ancestors (Bourdillon, 1987: 70).

The power of traditional leaders is measured by their control of land. Bourdillon (1987: 69) observes that the chief's control of the land by virtue of his seniority, he intercedes with the ancestral spirits for the well-being of the chiefdom. Land therefore is the *sine qua non* of traditional leadership authority. The allegiance of subjects to traditional leaders was premised on the latter's ability to provide land that would ensure security in old age (Bourdillon, 1987). During the coronation of Chief Ziki (a Karanga chief) according to Bourdillon:

> ...we are told that the new chief opened his hands that were filled with two handfuls of earth while he was addressed: 'You are now Ziki. We hand you the country to hold. Look after us well'. Similarly, we find in the history of the *Va*Shawasha people that when their chief had occupied the country and killed the old Rozvi chief, he called all his

people together; when they gathered, one of his followers dug a handful, mixed it with water and gave it to him saying, 'Take this soil Chinhamora.' When he took the soil all the people clapped in recognition of his chieftainship (Bourdillon, 1987: 68).

In Shona cosmology, marriage is regulated by taboos; for instance, the month of November is sacred. Traditionally, the Shona people would not marry during that month because they respected the fertility of the land. This observation is based on the appreciation of the fertility of the land, which has both a metaphorical and a direct relationship with the fertility of households. In Shona everyday discourse, soil is equated to women: when a man showed interest in a woman he would say, *"Gombo iri ndarida ndinoririma chete"* (I love this woman, and I will propose to her). The *gombo*, a piece of land that has been lying fallow, is equated to a woman. The fact that the *gombo* has been laying fallow means that it is fertile and if crops are planted in it, they will be healthy. Women like land are equated with fertility and land directly affects the fertility (and the livelihood) of people. A woman represents the soil/land and a man represents the water. In as much as land is fertilised by the water/rain, so is a woman by a man. In this symbolism, households are dependent on the reproduction of nature. In other words, harmony and viability in the ecologically married soil and water translate into the harmony of the human marriage. It is this harmonious relationship between water and soil that translates into a holy matrimony.

Musiyiwa (forthcoming) suggests that land must perform its fertility first and if land was not fertile people would not be fertile. In addition to this, the soil is conceptualised as the mother because people are buried in the soil, so the soil looks after those buried in it just like a mother looks after her children (see also Musiyiwa, forthcoming). People grow crops from the soil, and livestock graze on that which grows in the soil. Soil fertility is intimately linked to ancestors and this corroborates why ancestors impose taboos to maintain the fertility of the soil further illustrating why the Shona would consider the marriage between water and soil as a holy matrimony. Therefore, when rain falls, the water activates the fertility of the soil.

In Shona thought, water and soil are very valuable assets as they regulate the rhythm of life, both annually and in a lifetime. The land is pivotal in the lives of the Shona be they in the rural or urban areas. Bourdillon says:

> The land as a productive resource remains of crucial importance to the Shona even today, whether it be for subsistence during life or simply for security in old age…The land is thus important for the continuity of a people with its traditions. The land links past and present, the dead and the living, the chief and his people, and it binds people together (Bourdillon, 1987: 71).

The intimacy between the Shona and the land is vividly expressed by an incident I witnessed at Msipane Business Centre in rural Zvishavane, during my fieldwork in 2012. An elderly male passenger, a local person in the Msipane area, disembarked from a bus travelling from Zvishavane town. Apparently, he quarrelled with the bus conductor and when he disembarked, he challenged the bus conductor to a fistfight. He declared, *"Mfana chiuya tichigwa manje nokuti ndave muvhu rekwedu"* (Young man, we can now fight because I am now in the land of my ancestors). Further illustrating the bond between the Shona and the soil, Bourdillon cites a Shona story:

> *Mwari* (God) distributes goods to men: to one man he gave people and cattle and to another only a handful of soil; the latter was able to claim as his own all that grew in or fell on the soil, including all the people and cattle born on the land. Through possession of the land he became a great chief while the man who received followers and cattle became a headman under the owner of the soil (Bourdillon, 1987: 67).

The analysis above did not escape Mr Phiri Maseko's conceptualisation of water and soil. His assertion that he "marries water and soil" resonates with the larger Shona conceptualisation. However, in his conceptualisation of water and soil, Mr Phiri Maseko goes a step further in his formulation that extends to the practical and the infrastructural: for the marriage of water and soil to be productive, the water must be harnessed because, if left to flow, the water will 'elope' with the soil. If the water and soil elope, he would not be able to produce much on his plot. Moreover, if the water and soil were to elope, one of the deleterious results would be soil erosion leading to gully formation. Thus, the marriage (water and soil in holy matrimony) ensures ecological harmony on his plot in a very practical way for smallholders.

The travails of smallholder farmers

The term smallholder farmer is used interchangeability with terms like subsistence, small-scale, resource poor and to a less extent, peasant. However, the term *peasant* appears to have disappeared from the lexicon of smallholder agriculture. This book will use the term smallholder farmer as there is no consensus on which indicator to use when defining the term smallholder (Aidenvironment, 2013: 11). The following Aidenvironment (see Table 3) of indicators would be useful in defining a smallholder:

Table 3 suggests that the definition of the term smallholder farmer is problematic. Cousins agrees:

> I argue that the term (smallholder) is problematic because it tends to obscure inequalities and significant class-based differences within the large population of households engaged in agricultural production on a relatively small scale. Much usage suggests that smallholders form a relatively homogenous group, and fails to distinguish between producers for whom:
> • Farming constitutes only a partial contribution to their social reproduction
> • Farming meets most of their social reproduction requirements
> • Farming produces a significant surplus allowing profits to be reinvested and, for some, capital accumulation in agriculture (Cousins, 2010: 3).

Cousins (2010) implies therefore that characteristics that constitute smallholder farming contextually vary.

Landholding size	Surface of cultivated area (either a number in hectares (ha) or relative size in comparison to regional/national or sector average). Location (rural, isolated versus urban).
Labour input	Ratio family labour versus hired labour. Permanent staff or seasonal labour. Amount of labour input. Distance between residence and farm.
Farm management responsibility	*Income* Share household income from farming Multiple income sources (off-farm activities) Costs of an audit compared to commodity value
Farming System	Level of technology Irrigation versus rain-fed
Capacity	Farm management Administration Communication in the language of certifier Storage, processing and marketing Certification
Legal aspects	Registered as a private company Land tenure

Table 3: List of indicators to define smallholder (Source: Aidenvironment, 2013:11)

Smallholders in rural Zimbabwe mostly depend on rain-fed agriculture and to a very limited extent on irrigation. Thus, their agricultural livelihoods are susceptible to drought. Smallholder farming is characterised by low levels of mechanisation and the farms are generally small and held under complex tenure that is traditional or informal. Smallholders produce for subsistence and commercial purposes. Many are motivated by the need to replenish the strategic grain reserve for the community thus ensuring that there is food available for all. Such smallholders demonstrate the *hurudza* concept of Shona societies. This term *hurudza* is explored further in Chapter 2.

The smallholders depend on family labour while the more economically privileged occasionally make use of hired labour. At times, the smallholders revert to the communal concept of *humwe* (work parties) to complete the various chores during the agricultural season. During such occasions, villagers come together and help till the land at each other's plots.

The smallholders' plots are characterised by poor soils thereby exacerbating smallholder farmers' food insecurity. Food security in Zimbabwe is primarily based on maize production; however, small grains ('orphan crops') such as millet and sorghum play a peripheral role (Jayne et al., 2006). About three quarters of smallholder farmers live in areas with less than 650mm of rainfall per annum (Jayne et al., 2006).

Rainfall variability is one of the most enduring threats to smallholders' livelihoods in semi-arid Zimbabwe. The World Bank Independent Evaluation Group (IEG) (2012: 31) says, "rain-fed agriculture is sensitive to climate variability – too little or too much rainfall...this is especially true in the dry lands... home to many of the poor, and many dependent on agriculture". Population increase and high unemployment have increased pressure on agriculture to provide food; rainwater harvesting offers a viable option for increase in food production for agriculturally dependent communities (Everson et al., 2011: 1). Moreover, with increased uncertainty, water harvesting offers a mechanism for adapting agriculture to climate change (Nicol et al., 2015). Nicol et al., further suggest that there is need of capture of rainwater and an optimisation of managing rainfall through water harvesting can lead to the improvement of smallholders' food security by upgrading rain-fed agricultural production. Nicol, et al say:

Water harvesting systems have the following characteristics: overland flow harvested from short catchment length, catchment length usually between 1 and 30 cm, run-off stored in soil profile...the general design principle of water harvesting involves a catchment area that collects run-off coming from roofs or ground surfaces...Rainwater harvesting provides many advantages over other irrigation schemes. They are simple and inexpensive to construct and can be built rapidly using local materials and manpower...techniques include pitting, earth basins, semi-circular bunds, contour ridges (Nicol et al., 2015: 133-134).

The preservation of water through harvesting is critical if farmers are to build resilience to climate variability. Research in South Africa that was conducted between 1993 and 2005, proved that water harvesting increased infiltration of water in KwaZulu-Natal with contour bunds thereby reducing irrigation needs by as much as 50% (Auerbach cited in Auerbach, 2013). A research in the Free State province also confirmed the importance of water harvesting and its potential to reduce risk of crop failure (Botha et al., cited in Auerbach, 2013). Auerbach (2013) says that rain-fed farming in arid areas is effective with a combination of organic soil management that increases the soil's capacity to increase water and nutrients, and water harvesting techniques and their potential to reduce crop failure due to variable rainfall.

The Africa Agriculture Status Report (2014) produced by AGRA warns of the dire effects of "failed seasons" on smallholder livelihoods due to climate change. The report of the Intergovernmental Panel on Climate Change (IPCC) (2007) suggests that a 0.76 degrees Celsius increase in the world's average temperature will slow, or even reverse human development, further compounding deep national and global inequalities. There is general agreement among scientists that there will be an increase in the frequency of weather hazards, less predictability of the variability of climate and a degree of uncertainty, in particular the possibility of rainfall scarcity and its impact on the livelihoods of smallholders who depend on rain-fed agriculture (Boyd and Cornforth, 2013).Given that climate change is already taking its toll on smallholder farmers and that it will inevitably have progressively more adverse impacts in the future, building resilience to these changes has become a rallying cry for action (AGRA, 2014). Tales of how resilience is built should be told as Baldwin below says:

While the tale of how we suffer, and how we are delighted, and how we may triumph is never new, it must always be heard (Baldwin, 1965).

Similarly, in the prologue to his book, *The seed is mine – the life of Kas Maine, a South Africa sharecropper*, van Onselen writes:

A historian's foremost duty is to ensure that merit is recorded and to confront evil deeds and words with the fear of posterity's denunciations (van Onselen 1996: xi).

This dictum inspired this study to attempt to tell the life history of Mr Phiri Maseko. I document his story to draw lessons from his agricultural practices; this, to rethink existing strategies for adaptation to climate change. I suggest that narratives of enterprising smallholder farmers can provide crucial insight into their experiences, and help to inform the management of climate variability in rural Zimbabwe. This will also help to fill the gaping hole left by the effects of the hegemonic developmental paradigm with its attendant linear model. Critchley et al., suggest that smallholder farmers have the potential to:

comprise a 'storehouse' of existing knowledge and ideas;
provide a fast track towards successful and adoptable land husbandry systems;
provide a direct and quick entry into a community;
constitute a 'pre-selected' team with which to work;
respond well to recognition (through the psychological mechanisms of 'positive feedback' and 'reinforcement');
network well together;
make good on-farm researchers (as they already have relevant experience and inquiring minds); and
enjoy spreading knowledge (in many cases) (Critchley et al., 1999: 14).

Smallholders employ a suite of strategies to respond to threats. Hastrup (2009) focuses on how humans employ strategies to reshape their livelihoods in response to threats; she suggests that it is vital to address climate-related threats from the perspective of the people living with the threat. To that end, this study attempts to address responses from 'below' from the perspective of Mr Phiri Maseko – through his innovations as he tries to build resilience in the face of climate variability. Water-Bayers and van Veldhuizen (undated) assert that sharing of local innovations can provide ideas and inspiration for others to experiment with and adapt new ideas to their own setting. The work of a multi-stakeholder global partnership programme for research and development, Promoting Local Innovations in Ecologically Oriented Agriculture and Natural Resource Management (Prolinnova), inspired me to try to undertake an ethnographic study of the efficacy of innovations by Mr Phiri Maseko and his adopters in rural Zimbabwe. Prolinnova shares lessons and builds on the experiences of smallholders to promote and support their innovative capacity through farmer-led joint research and development. Prolinnova's goal is "to develop and institutionalise partnerships and methodologies that promote processes of local innovation for environmentally sound use of natural resources" (Farmer-led Documentation 2012: 2).

Such local innovations can be understood through ethnography, a data generating method mainly used in the field of Anthropology. The discipline of Anthropology offers a broad panorama of the lived realities of research participants and has the potential to help unravel the complex and nuanced processes and contexts that contribute to adaptation to climate. Anthropologists therefore "should not be too bashful and seek to hide our lights under a bushel" to borrow an expression from Marwick and Emsley (1989: 4). A sub-discipline of anthropology - agricultural anthropology - a term popularised by Robert Rhoades (cited in Veteto and Crane, 2014: 1), offers more insights about smallholders and is "broader in scope than other agricultural disciplines which focus, and rightly so, on specialised and limited problems in agriculture". Ethnography, a sustained observation, was insightful in deciphering smallholder innovations in Zvishavane.

Zvishavane district is susceptible to severe droughts, which have warranted a multiplicity of responses to climate related risks making it an ideal site to study innovative strategies to a changing climatic environment. The existence of traditional structures, new institutions, new spaces of power and power configurations emanating from the country's political landscape in a terrain dominated by Non-Governmental Organisations (NGOs) (often not sharing state ideology) makes Zvishavane a suitable district to investigate interventions and, contestations over knowledge generation, dissemination and use. Mr Phiri Maseko's project is ideal to study in the background of these exogenous interventions. This study attempts to chronicle the journey that Mr Phiri Maseko travelled. Admittedly, his project has been influenced by external sources of knowledge, but it has proved to be more resilient than donor driven interventions.

Donor driven projects appear to have short lifespans. In her PhD thesis, Mangoma (2011) writes about the *Ngwarati* Wetland-Tillage System in the dry Shurugwi rural area of Zimbabwe (which is adjacent to Zvishavane). The project from 1998 to 2002 was donor funded. The *Ngwarati* system is characterised by water harvesting and introduced by the Agricultural Technical and Extension Service (AGRITEX) in Zimbabwe's rural areas. The *Ngwarati* system "borrows from the traditional cultivation practice of ridges and furrows but reoriented at zero gradient and broadened to allow for the interception and retention of water and silt within the cultivated areas" (Owen cited in Mangoma, 2011: 13). The donor's withdrawal precipitated the collapse of the project, and its demise characterises projects that are top-down approaches. NGOs appear to be accountable to their funders and not aid recipients. Birdsall (2011: 22) points out that aid recipients may appear to be in the driver's seat, but they often have "donor passengers grabbing the steering wheel or shouting instructions from the backseat".

An NGO introduced Conservation Agriculture (CA) in southern Zimbabwe as a mode of adapting to climate by smallholder farmers. The principle of CA is the preservation of moisture and minimum soil disturbance. CA as practised in Zvishavane is very labour intensive because it is mainly manual. From my fieldwork, where the NGO withdrew, it appeared there was no exit strategy in place to ensure continuity of practice. These interventions contrast with Mr Phiri Maseko's brainchild project, which has potential to improve rural livelihoods. His agricultural practices can help boost smallholder agriculture. There is need to probe what motivates smallholder innovations. However, to date very limited work has been done on individual smallholders' motivation to innovate (Goldman, 1993; Li et al., 2000; Hall et al., 2003).

Mr Phiri Maseko's responses to climate and its variability are dependent partly on local ethno-ecologies, thus justifying an investigation into his knowledge generation and dissemination and how both the NGO community and the rural people (this is examined in Chapter 5) have used it. Rural people in Zvishavane have developed specific innovative systems. The challenges farmers face are many: inadequate resources, infertile soils and a variety of chronic crop-affecting diseases. The changing climatic environment adds to the list of challenges they face. This study refers to Mr Phiri Maseko's methods as the 'Phiri Maseko ecology'. His ecological approach is an example of an ethno-ecology at a micro-level, an attempt to try to address these problems faced by smallholders.

Leach et al., (2010) argue that making science and technology work for the poor such as smallholder farmers has become one of the major challenges in modern times, yet despite international aid at attempting to alleviate rural poverty, interventions often fail. Hillier and Castillo (2013) note that in many places of recurrent crises, international donors and governments' response are not good enough. They conclude that a new focus that builds resilience by smallholders offers real promise to the rural poor to thrive during shocks, stresses, and uncertainty.

A major hurdle is to help poor resource farmers secure better livelihoods that are also sustainable from their "complex, diverse and risk-prone farming when normal agricultural research has so largely failed" (Chambers, 1989: 195). This is particularly evident in the era of complexity. Hawkins (cited in Preiser, 2014) predicted that the twenty-first century would be the century of complexity.

O'Brien argues:

> Outcomes within complex systems are difficult to predict with certainty because small changes can have large consequences...This raises challenges for social responses, including adaptation, particularly if complex, nonlinear problems are addressed in a discrete, linear manner. Complex problems do not always act in ways that are expected, despite human efforts to consider all types of contingencies. While the notion of surprise is always relative to the viewpoint from which it is considered, there is little doubt that global environmental changes at the scale, rate and magnitude that they are now occurring, will lead to new and unexpected outcomes (O'Brien, 2013: pages not numbered cited in Mabeza et al.,).

Therefore, one response to this state of uncertainty and complexity might be innovative adaptive strategies that are aimed at harvesting water. Water harvesting could be an ideal innovative adaptive strategy in semi-arid southern Zimbabwe where Wilson (2010) says, "rain comes rapidly and leaves rapidly" leading to food shortages and soil erosion. Solutions to variable environments such as southern Zimbabwe should consider local dynamics.

Climate change and adaptation appear to be debated and analysed through a global lens that tends to project universally homogenous solutions to climate change related risks of smallholders. Such framing of the climate change challenges precludes an appreciation of local realities and knowledge systems, which should be considered enhancing rather than constraining, global efforts at helping smallholders adapt and innovate in a changing climatic environment. Zimbabwe's National Environmental Policy (Government of Zimbabwe, 2009) posits that local knowledge and traditional practices have an important role to play in the management and sustainable use of natural resources. However, it seems the authorities only pay lip service to the important role local knowledge plays in smallholder farmer livelihoods. Smallholders have intimate knowledge of the conditions and dynamics of their local environments more than the 'experts' (Roling, 2009). Therefore, new approaches that prioritise the rural poor in climate change agenda could tap into and activate the latent energy and talents of the innovative rural communities to bring about development that is sustainable.

One such latent source of energy is arguably, rural people's climate-related experiences, competencies and knowledge systems. Rural communities should be at the center of the innovation process as they have superior knowledge of their production processes and social context. An approach premised on the complementary role of local narratives plays a pivotal role in the process of understanding resilience among the rural poor. This approach positions the *last* (rural poor) *first* (Chambers, 1983). Interventions that address smallholder farmers' vulnerability to climate need to build on local innovations and local capacities. Conventional approaches in rural development replicate the top-down method of conventional development embodied in modernisation theory. Science, in the hands of the state and civil society has become a straightjacket in which rural communities must fit in matters of environmental management.

It is essential to gain an understanding of the factors driving smallholder adaptation and innovation (or lack of it) in the context of climate variability, natural resource governance and traditional/cultural practices. Adaptation must be considered to be a process in which rural communities gain access to vital skills, information and resources in order to continuously shape and adapt their lives and livelihoods in their situations. Understanding smallholders' innovations and adaptations in the 'tradition culture/climate change/natural resource governance' nexus, remains a research gap in Zimbabwe, and southern Africa. Yet, such an understanding has implications for climate change-related decision making at micro, meso, macro and supranational levels.

From my interviews during my fieldwork, southern Zimbabwe is home to innovations by smallholder farmers, which date back to precolonial times. Innovations by smallholder farmers have potential to play a pivotal role in rural development, especially in the wake of managing climate variability. Nonetheless, smallholder innovations appear to be the Cinderella of developmental policies with preference being given to the linear transfer of technology model initiated by research institutions. The limitations of conventional research and development (R&D) led innovation, which attempts to "develop technologies 'on behalf of'... are becoming clear. Externally developed technologies and practices are often not appropriate, and uptake is frequently very poor" (Letty and Bell, 2012: 1). Moreover, interventions for managing climate variability by NGOs appear to be one-size fits all solutions. Such solutions assume homogeneity among smallholder farmers and their social ecological contexts. It is this interventionist type of development that might be contributing to the insecure livelihoods of the smallholders. There is a need to bring on board local innovations as a way of addressing the impacts of unreliable rainfall. The smallholders in rural Zimbabwe also must contend with poor extension services compounded by the country's economy, which appears to be in a free fall. Soil fertility is in decline and smallholders do not have adequate financial resources to buy inorganic fertiliser.

Zimbabwe's agricultural sector was dualistic with thriving and well supported commercial farming (located in areas of rich soils and high rainfall amounts) and smallholder farming where people are in the throes of poverty (Interview with AGRITEX officer, 2012). Government policies from the colonial era were heavily biased towards commercial farming. Even after the country's independence in 1980, policies continued to favour interests of commercial farmers. Moreover, Letty and Bell (2012) observe that most the rural poor depend on small farms less than 1 hectare that are mostly located on impoverished land with very limited resources and bad infrastructure. Smallholder farmers in Zvishavane rural are no exception. Due to the inequitable colonial land policy in Zimbabwe, most smallholders were moved to reserves.

Addressing these challenges requires a 'back to basics' approach – a return to the concept of local innovations. The concept of local innovations has withstood the test of time and is also dynamic in nature.

Book outlook

Chapter 1 considers the contestation of adaptation to climate. It also cautions against the doomsday approach to climate change. The chapter considers the concept of vulnerability and its related concepts of exposure, sensitivity and adaptive capacity and how these concepts speak to resilience. The chapter employs literature from a resilience perspective to climate adaptation and traces the trajectory of adaptation as a concept and also relies on the literature on smallholder innovations to climate variability and gives an appraisal of smallholder innovations in East Africa. This study builds on the concept of barriers and enablers to a food secure pathway.

In Chapter 2, I situate the genesis of Mr Phiri Maseko's innovation in the "sweep of history" as Wilson would say. I argue in this study that the story of Mr Phiri Maseko of the marriage of water and soil would be incomplete if consideration is not given to what inspired his innovations for managing climate variability. There has been very little history written about his innovative methods trajectory. The chapter presents a multi-narrative approach to documenting the life of Mr Phiri Maseko to tell a nuanced account of his creative agricultural practices. This book makes a compelling case that if innovations by smallholder farmers are to play a more meaningful role in rural development then consideration should be given to what inspire these innovators. In other words, what inspires innovators could also help in turn to inspire other smallholder farmers to innovate. Towards that goal, this study documents the life story of Mr Phiri Maseko and suggests that his innovations are a marriage of both endogenous and exogenous knowledge. This chapter advances the argument that he encapsulated the Shona *hurudza* concept.

After situating Mr Phiri Maseko, Chapter 3 takes the book to the heart of the matter, the 'marriage of water and soil' the innovations for managing climate variability. I begin the chapter by making an analysis of the Phiri Maseko ecology. This serves as a prologue to Mr Phiri Maseko's marriage of water and soil as demonstrated by his innovations. I argue that the Shona conceptualisation of water and soil is at the heart of the 'Phiri Maseko ecology'.

In Chapter 4 this book investigates the efficacy of Mr Phiri Maseko's innovations in other areas of Zvishavane where they have been replicated, what I call 'rhyming with an audience'. I argue that there is also great creativity within the ranks of Mr Phiri Maseko's adopters. Data evidences a high uptake of his agricultural practices because he acts in the same metaphoric/analogous realm of the conceptualisation of water and soil as other farmers. The chapter examines the barriers and limits to adoption of innovations.

The Conclusion is a clarion call for what the study calls "good news" makers to stand and be counted. In other words, the study makes a compelling argument that despite the doom and gloom pervading society today, success stories of farmers such as Mr Phiri Maseko give us hope about human capacity to adapt to climate. I argue that through his innovations that catapult him to a *hurudza,* Mr Phiri Maseko redefined adaptation to climate and thus becomes a *good* news maker. The chapter analyses the dominant narrative with its emphasis on the linear transfer of technology and the narrative of smallholders as needing external interventions. I also suggest that in this age of uncertainty due to increasing climate variability, replicating Mr Phiri Maseko's innovations offers food security in semi-arid rural Zimbabwe.

Chapter 1

Full of sound and fury?
The climate change discourse

> We are now faced with the fact that tomorrow is today. We are
> confronted with the fierce urgency of now. In this unfolding
> conundrum of life and history, there "is" such a thing as being too
> late. This is no time for apathy or complacency. This is a time for
> vigorous and positive action." — Martin Luther King Jr. (1963)

"Post-truth age": The new normal?

This chapter foregrounds the study on its theoretical perspectives
based on the contested perspectives in the "crowded and noisy"
world of climate adaptation as O'Riordan (cited in Hulme, 2009)
would say. The research grapples with the concepts of adaptation,
resilience, vulnerability and innovation. The study uses the theoretical
perspectives lens to try and identify barriers and enablers to a food
secure pathway. Unless there is an understanding of what constitutes
hurdles to adaptation, the pathway to food security will be fraught
with huge challenges. Thus, this chapter begins by challenging the
doomsday approach, that humanity is facing "imminent peril". This
study sees the doomsday approach as a hurdle because as Stoknes
(2015) argues, it leads to "greater resistance, making it harder to enact
measures to reduce greenhouse emissions and prepare communities
for the inevitable change ahead".

In as much as the study acknowledges the threat posed by climate
change and the need for immediate action, the "fierce urgency of
now", wallowing in the doomsday approach is counter-productive.
The doomsday approach also leads to what Stoknes calls a kind of
"apocalypse fatigue", when more than 80% of media articles use an
apocalyptic narrative for climate change. Resultantly, this leads to the
consumers of the information becoming tired of the doomsday
narratives and stereotyping the messengers negatively (Stoknes,

2016). This must be countered by urgent action to correct this misconception.

In 1963, at the March on Washington, Martin Luther King Jr articulated what he called "the fierce urgency of now", the need for urgent action on civil rights (Keohane, 2015). Nowhere is this relevant today than the threat of climate change. Thus, the fierce urgency of now is the need to change our perceptions, the way we frame climate change. Instead, there is need to consider Mike Hulme (2009)'s advice that we build on opportunities that climate change present to us. Such an advice should make us pause and ask: Are we witnessing a planetary march to Armageddon as doom mongers remind us *ad nauseam*? Has the climate change doomsday clock struck five minutes to midnight? Is the dystopic feeling that permeates our lives the work of false prophets on the prowl? We are warned in the tradition of the prophets of old that we will perish.

Has humanity entered the post-truth age? 'Post-truth' was chosen the word of the year 2016 by Oxford dictionaries (BBC News, 2016). The Oxford dictionaries define 'post-truth' as an adjective "relating to circumstances in which objective facts are less influential in shaping public opinion than emotional appeals". Grathwohl of Oxford Dictionaries says that social media with its distrust of facts offered by the establishment, engendered the etymology of 'post-truth'.

Zahira Jaser, argues that the 'post-truth' modifies the way we make sense of the world. She observes that effective leaders adapt their style to different contexts. For example, in the post-truth context, political leaders have a capacity to create a reality, neither true nor false but meant to stir emotions and beliefs of society (Jaser, 2016). With the recent events, Brexit (British exit from the European Union) and the rise of Donald Trump in the United States of America (USA), perhaps the "post-truth age" is now a reality. In both cases (Brexit and USA elections), the winners appeared to stoke their bases by "cherry-picking" information that would be sweet music in the ears of their audience (Jaser, 2016). Re-phrasing Sullivan and Smith (2016)'s account of how Fidel Castro was accepted as a fact of life in Cuba, this study suggests that the "post-truth" is now with us, like climate change – what good would it do to doubt its advent?

Taking a cue from the definition of post-truth, this study argues that in as much as climate change poses a grave threat to humanity, the doomsday approach appeals to subjective emotions rather than objective facts and thus, fails to appreciate the potential for human ingenuity in addressing adversity. Appealing to our emotions as encapsulated in the doomsday approach means that ultimately, climate change is interpreted differently according to our interests. Livingstone puts it eloquently: "debates about climate change turn out to be disputes about ourselves – our hopes, our fears, our aspirations, our identity".

Whatever the case might be, we seem to have entered the Anthropocene, a planetary era characterised by human induced climate change. The term Anthropocene, however, has been fodder for critics who view it as too anthropocentric and misleadingly general in scope, too keen on evidence of Man and of "our" collective imprint on the globe, to the exclusion of profound differences of responsibility and vulnerability regarding contemporary ecological crises (Howe and Pandian, 2016). Critics further argue that declaring an epoch in our name is brazen evidence of an ultimate act of "apex species self-aggrandizement—less a geological -cene, perhaps, than an Anthropo*scene*" (Pandian, 2015). In as much as these critics shout from the rooftops about their discomfort with the term, surely, we cannot discount the implications of the negative effects that humans have had on the environment.

Eugene Stoermer coined the term Anthropocene in the 1980s and it went on to be popularised by Nobel Laureate Paul Crutzen (Stromberg, 2013). Anthropocene "has become something of a buzzword as it enters its mid-teens" (Castree, 2014: 234). Levene (cited in Castree, 2014: 234-235) suggests that the Anthropocene concept is yet to "become standard currency, though there has been sufficient acclamation from a wide range of scientific and non-scientific disciplines to suggest its durability". That it (the Anthropocene) has not become standard currency only gestures towards the contestation of its genesis. Debates on the authenticity of anthropogenic climate change have been heated, what this study calls a discussion full of sound and fury as Shakespeare would say. However, this should not be misconstrued to mean that humans have been let off the hook as alluded to above.

The discussion on the Anthropocene taps into the tradition of the memento mori, to borrow a phrase from Richard Noble. Humans are said to have caused extinction of plant and animal species, pollution of the oceans and altered the atmosphere. By altering the atmosphere, human activities have led to anthropogenic climate change that has committed the earth to global warming likely to exceed the two-degree Celsius threshold regardless of mitigation efforts (Berrang-Ford et al., 2010). With the never-ending policy stalemate of the major greenhouse gas (GHG) emitters shown by failure to establish a framework for emissions stabilisation, four degrees Celsius of global warming is likely by 2100 (Parry et al., cited in Berrang-Ford et al., 2010). The sections below consider the theoretical perspectives of the study and this further affirms Hulme's assertion that climate change means different things to different people in different locations. This assertion means that because of these differences, climate change is about interests, thus, becoming a high stakes game characterised by sound and fury. The theoretical perspectives bare testimony to the sound and fury that characterise disagreements in the climate change debates. Nevertheless, climate change has become a reality according to many scientists.

The effects of climate change are being felt today. Exploring ways of living with these effects merits urgency and is important for human development (Pelling, 2011). It is imperative to understand how to help the most vulnerable become food secure given expected changes to climate variability (Ziervogel et al., 2006). This book explores Mr Phiri Maseko's adaptation to climate through the lens of resilience and how it affects adaptation. Underpinnings of adaptation together with the interrelated concepts of vulnerability, resilience and innovation are explored. These concepts belong to different research communities and thus cannot be exhaustively discussed in this book. These concepts are intertwined and will be approached as such. The study will use vulnerability insights to explore why Mr Phiri Maseko

4

built resilience (thus becoming food secure to shocks and stresses. In this study, I use a multi-stressor approach (O'Brien et al., 2009) as a way of addressing vulnerability and ultimately food insecurity in Zvishavane. Mushongah (2009: 304) argues that, "one lens to understanding vulnerability is not enough". Vulnerability is affected by multiple environmental and social processes (Reid and Vogel, 2006; O'Brien et al., 2009; Shackleton and Shackleton, 2012). Thus, this book also draws on the concept of innovation as encapsulated by Mr Phiri Maseko and smallholders who have replicated his adaptive strategies to build resilience to a multi-stressor environment. It is within the scope of this book to probe the genesis of his innovative preparedness to climate variability to fully grasp the essence of his food security. Further to that, I also look at how and why his innovations have spread elsewhere in Zvishavane.

The need to develop resilience in society and ecosystems is considered as a cornerstone of adaptation to change and uncertainty (Adger et al., 2009). Because of the realities of climate change in Africa, sustainability and resilience are likely to play a pivotal role for the sustenance of smallholder farmer livelihoods (Auerbach, 2013). Knowledge on how to strengthen the resilience of a system in both society and nature or social ecological systems is considered important when dealing with climate change impacts (Moberg/Stockholm Resilience Centre, 2014). Pelling (2011) reinforces this argument by explaining that resilience is important in systems experiencing extreme weather and unpredictability. Meybeck et al., (2012) argue that response to food crises by governments and international donor organisations has been inadequate and consequently, a new approach premised on building resilience promises to help communities respond to stresses and shocks.

Smallholder farmers are vulnerable to the interplay of many stressors hence the need to build resilience. Shackleton and Shackleton (2012: 276) argue that "in considering vulnerability, it is also necessary to define resilience, as building resilience is key to reducing vulnerability and enhancing adaptive capacity". Folke defines resilience as:

...a fundamental concept in complex social ecological systems thinking and refers to the situation where social ecological systems, households or communities are able to respond to shocks and stresses and, moreover, use this as an opportunity for innovation and adaptation (Folke cited in Shackleton and Shackleton, 2012: 276).

Vulnerability is "the degree to which an individual, group or system is susceptible to harm due to exposure to a hazard or stress, and the (in) ability to cope, recover, or fundamentally adapt (become a new system or become extinct)" (Tompkins cited in Levina and Tirpark, 2006: 17). Shackleton and Shackleton (2011: 275) argue that vulnerability is a "complex concept with several definitions and applications across disciplines and contexts". Kelly and Adger cited in Shackleton and Shackleton (2012) say vulnerability encompasses issues related to risk or shocks as influenced by varied environmental and social processes. According to the IPCC cited in Shackleton and Shackleton (2012: 276), vulnerability has three dimensions as follows: (i) exposure, (ii) sensitivity, and (iii) capacity to respond, or adaptability (see Table 1).

Adger (2006) emphasises that the stresses an individual, group or system is exposed to, is associated with environmental and social change. Although the use of the term vulnerability has its roots in geography and natural hazards research, the term is now a central concept in such research contexts as climate impacts and adaptation (Hans-Martin, 2006). Downing (cited in Brooks, 2003) observes that researchers from the natural hazards field often focus on the concept of risk, while those from climate change and social sciences opt to focus on vulnerability. Climate scientists regard vulnerability as the likelihood occurrence of weather and climate related effects while on the other hand social scientists view vulnerability as representing socio-economic factors that shape people's ability to cope with stress or change (Allen cited in Brooks, 2003). Wisner et al., (2004) in justifying why vulnerability matters when building resilience to disasters, argue that analysing perturbations allows us to show why they should *not* be separated from everyday living.

Term	Characteristics	
Exposure	Largely deals with the degree of exposure to external hazards and shocks such as floods, droughts, wars, disease etc.	
Sensitivity	Indicates how at-risk a household or community is to the shocks and the likely degree of impact and change associated with these. It is largely defined through the assets people have, the activities and livelihoods strategies they adopt (such as agriculture). Various internal social, economic and structurally related factors.	
Adaptive capacity	Relates to the ability to deal with and recover from exposure to a shock or stress. Vulnerable households and communities typically lack this capacity. This situation is sometimes referred to as 'defencelessness' and could be thought of as function of high sensitivity and low resilience.	

Table 1: The three dimensions to vulnerability (Adopted from Shackleton and Shackleton, 2012)

After defining the above terms, I consider the concept of resilience in more detail. The term 'resilience' gained currency in the 1970s through the pioneering work of C S Holling in the field of ecology, as a concept meant to help comprehend the capacity of ecosystems with alternative attractors to persist in its original state subject to disturbances (Folke et al., 2010; Pisano, 2012). However, resilience has come to mean different things because of the way it has been employed by different disciplines (UNDP, 2014). Walker and Salt (2006: 1) define resilience from an ecological perspective, as "the ability of a system to absorb disturbances and still retain its basic function and structure". The concept of resilience facilitates a more robust understanding of the interaction within systems (UNDP, 2014). There has been a call for innovations that can help build resilience and at the same time foster sustainability often referred to as social-ecological innovations (Moberg/Stockholm Resilience Centre, 2014).

Little and McPeak (2014) argue that resilience as a concept has been used in development thinking and practice for many years but its resurgence can be attributed to two topical issues. Firstly, the term resilience has gained traction because of the advent of climate change adaptation with its emphasis on building community resilience in the wake of an increase in extreme weather events. Secondly, because of food insecurity, especially that closely related to drought and conflict in the Horn of Africa (Little and McPeak, 2014). The call to "end hunger" by international donor agencies like the US Agency for International Development implied building resilience and this has been instrumental in "propelling the concept of resilience" to centre stage in development planning and actions (USAID cited in Little and McPeak, 2014). Resilience is increasingly gaining recognition and the concept has come to be viewed through a climate change lens (Vogel et al., 2007). In this study, I probe the concept of resilience from a perspective of how as a water harvester, Mr Phiri Maseko improved his food security in response to vulnerability.

Building resilience is about examining smallholders' vulnerability to a multi-stressor environment. An appreciation of one's environment informs how one responds to shocks and stresses. This study probes the role of values and beliefs in Mr Phiri Maseko's response to stresses in a variable environment.

Values and beliefs are central to culture and influence the way people build resilience as a way of adapting to dramatic environmental conditions such as climate change (Roncoli et al., 2009).

Roncoli et al. argue that:

> ...culture frames the way people perceive, understand, experience, and respond to key elements of the worlds which they live in. This framing is grounded in systems of meanings and relationships that mediate human engagements with natural phenomena and processes. This framing is particularly relevant to the study of climate change, which entails movement away from a known past, through an altered present, and toward an uncertain future, since what is recalled, recognised, or envisaged rests on cultural models and values (Roncoli et al., 2009: 87).

Ziervogel (2002: 275) adds that: "Rural livelihoods are complex bundles of survival strategies embedded with cultural preferences and social norms". In other words, how individuals and communities adapt is framed on common ideas of what is "believable, desirable, feasible, and acceptable" (Nazarea-Sandoval cited in Roncoli et al., 2009: 87).

This study tackles perceptions about factors that motivate building resilience to changing climate by smallholder farmers in Zvishavane. O'Brien (2012: 307) suggests that complex issues such as climate change require a change of perceptions, what she calls "adaptation form the inside-out". The following questions are pertinent to this investigation: Are Mr Phiri Maseko's perceptions for building resilience to climate appropriate for other smallholder farmers? Does his story make us re-think the concept of building resilience to climate? Do beliefs and values influence how smallholder farmers build resilience to climate? The study addresses adaptation through ethnography by drawing on the experiences of Mr Phiri Maseko and other smallholder farmers who improve their food security by replicating his farming practices in Zvishavane. The concept of adaptation is a subject of intense debate among scholars in the climate adaptation community.

Adaptation: The buzzword for today

To fulfil the mandate of this book, I will explore the dominant narratives that underpin climate adaptation interventions. I will show the shortcomings of some of the interventions and argue that adaptation means creating innovations without rejecting other forms of knowledge. In this study, I will account for the emergence of the term adaptation and its rise to prominence in climate change discourse.

Adaptation, 'long the poor cousin of mitigation,' has risen to become the pre-eminent item of climate science. Schipper and Burton (2009:1) posit that, "adaptation as a scientific concept was largely associated with the Darwinian theory of evolution and the process of natural selection". Over time, the term came to be associated with climate change. Adger et al., (2009: 2) note that adaptation to climate change is now part of the contemporary political and economic dialogue of global climate change. Adger et al., (2009: 2) further observe that adaptation has been included in the policy debate because of its appearance in Article 2 of the United Nations Framework Convention on Climate Change (UNFCCC), "where the ultimate objective of the Convention concedes that adaptation to climate change in relation to food production, ecosystem health and economic development can and will occur".

In the 1990s and 2000s, mitigation, dominated the international climate change dialogue, yet the past decade has witnessed the advent of adaptation, what I refer to as the 'new kid on the block' (e.g. Parry et al., 1998; Pielke et al., 2007 cited in Adger et al., 2009). After the term adaptation, had gained recognition, other terms such as coping, risk management, vulnerability reduction and resilience came into use (Schipper and Burton, 2009).

The realisation of the "inevitability of climate change based on the Inter-Governmental Panel projections regardless of reduction of GHG emissions, brought adaptation to the fore" (Pielke Jr, 2009: 352). Adaptation is significant given the rising levels of vulnerability and risk (Shackleton et al., 2013). However, this study agrees with scholars who argue that adaptation and mitigation are complementary to each other. Nevertheless, what *is* adaptation?

The term adaptation is contextual; it means different things to different people. For the purposes of my study, Moser and Ekstrom define adaptation in the following way:

> involves changes in socio-ecological systems in response to actual and expected impacts of climate change in the context of interacting non-climatic changes. Adaptation strategies and actions can range from short-term and long-term, deeper transformations, aim to meet more than climate change goals alone, and may or may not succeed in moderating harm or exploiting beneficial opportunities (Moser and Ekstrom, 2010: 1).

This book builds on Moser and Ekstrom's seminal definition. Moser and Ekstrom (2010) argue that adaptation strategies may succeed or fail contrary to assumptions that adaptation guarantees success. They also recognise the importance of meeting a range of goals beyond those of adapting to climate change alone. O'Brien agrees:

> ...adaptation responses to environmental change often seem to be constrained by the projections of climate models and integrated assessment models as if the future has already been decided and the challenge is for humans to adapt. A recent *Progress in Human Geography* Forum discussed the dangers of environmental determinism, including the risk of simplifying the causal drivers of global environmental change... (O'Brien, 2012: 668).

This is a realisation that adaptation to climate is subject to navigating many barriers some of which are not related to climate change alone. Adaptation is viewed as a continual process of adjustment and change instead of something that can be achieved through building up assets and capital only (Bernier and Muizen-Dick, 2014).

Pelling (2011: 13) speculates that the debate on climate change adaptation is driven by four questions: What to adapt to? Who or what adapts? How does adaptation occur? Moreover, are there limits to adaptation? Exploring the interplay of stressors among smallholder farmers in Zvishavane is an attempt to address these questions. This study identifies the types of adaptation that Mr Phiri Maseko deploys to build resilience. Do his adaptive strategies prevent loss, spread risk or engage in restoration? Do adaptive strategies by ZWP (an NGO Mr Phiri Maseko helped to establish) and other NGOs operating in Zvishavane, amount to 'adaptation by ribbon cutting'? The term 'adaptation by ribbon cutting' was popularised by Kay (2009). The term refers to decision-making biases regarding approaches to adaptation. 'Adaptation by ribbon cutting' is characterised by grand ceremonies amid much fanfare when an official (government or non-state) cuts a ribbon to officially open, for example, an infrastructure such as a dam. Kay (2009) notes that the cutting of the ribbon is a concrete sign of progress, of a decision made that has successfully led to a tangible outcome. Kay argues that 'adaptation by ribbon cutting' emphasises two characteristics of a form of decision-making:

- A focus on infrastructure construction as an adaption option, coastal engineering works (such as seawalls, groynes and breakwaters), water management works (irrigation channels, river dams, channel diversions), flood protection works (embankments, overflow channels), and so on. For example, the construction of a seawall is chosen in preference to building the resilience of coastal ecosystems (such as sand dunes or mangroves) or the relocation of assets and communities at risk from coastal erosion.
- A short-term decision frame in preference to long-term adaption decision making. The motivation behind the short-term focus being the hope from those sponsoring the adaptation infrastructure works that they will personally be present to cut the ribbon within their term of office and thereby in the public spotlight to reflect the success of this good deed (Kay, 2009).

In an article in *The Tiempo* (73, 11- 15) Kay concluded:

If we are to identify effective adaptive solutions, we must be prepared to think beyond short-term political horizons and outside the confines of self-interest and expediency. (Kay 2009: 15).

'Adaptation by ribbon cutting' prompts questions such as what climate actions are likely to boost food security and have adaptive strategies used in the past helped to prevent the erosion of livelihoods? There is emerging consensus that incremental adaptation (small adjustments to current responses) alone is not adequate to help ensure that climate impacts do not erode livelihoods (Howden et al., 2010; Pelling, 2011; Stafford-Smith et al., 2011; Park et al., 2012; Tschakert et al., 2014).Park et al., (2012: 119) recognise that the common feature of actions that characterise incremental adaptation "lies in their central aim of maintaining the essence and integrity of an incumbent system or process at a given scale". Transformative actions have been suggested as a way of helping address climate change impacts and reducing vulnerability (Ribot, 2011; Park et al., 2012; Ribot, 2014; Tschakert et al., 2014; Bahadur et al., 2015). The main difference between incremental and transformational change is the extent of change, in practice manifesting either in the maintenance of the present system or process (incremental), or in the creation of a fundamentally new system or process (transformational) (Park et al., 2012).

The evolution of the concept of transformation can be traced to the Intergovernmental Panel on Climate Change (IPCC)'s Fifth Assessment Report (AR5) (2014). AR5 offers three narratives of the concept of transformation:

- transformation inducing fundamental change through the scaling up of adaptation, conceived as a limited, technical intervention with transformative potential;
- transformations as sanctions or interventions opened when the limits of incremental adaptation have been reached;
- transformation seeking to address underlying failures of development, including increasing greenhouse gas emissions by linking adaptation, mitigation, and sustainable development.

13

Pelling et al., suggest that:

> Transformation as an adaptive response to climate change opens a range of novel policy options. Used to describe responses that produce non-linear changes in systems and ecological environments, transformation also raises distinct ethical and procedural questions for decision makers. Expanding adaptation to include transformation foregrounds questions of power and preferences that have so far been underdeveloped in adaptation theory and practice (Pelling, et al., 2014: 1)

There is a need to re-evaluate existing structures, institutions, habits, and priorities considering risks posed by climate change (Rickards, 2013). Transformational change has the potential to play a key role in shifting systems to help initiate more fundamental change to adapt society to adverse effects of climate change (Rickards, 2013). O'Brien (2012: 673) reinforces this argument about the need to re-evaluate the status quo. She argues that research in geography has focused largely on adapting to changes that are underway instead of focusing on research that helps to comprehend how to "deliberately transform systems and society in order to avoid the long-term negative consequences of environmental change".

Transformation actions include a change in land use or location, or increase in diversification of income generating strategies as well as a change in the scale at which the system functions (Howden et al., 2010; Park et al., 2012). Rickards and Howden (2010) write about two types of transformational agriculture (TA):

(i) Changes in goal (for example resulting in a major change in land use and/or something different; and /or

(ii) Changes in location (of an agricultural activity and/or agriculturalists) (Rickard and Howden, 2010: 243).

Rickards and Howden (2010) justify that both transformative actions mentioned above are aimed at minimising adopters' vulnerability. The two actions focus on reducing sensitivity and exposure respectively in the sense that where the anticipated change is negative a change in land use aims to reduce the adopters' sensitivity to impacts by changing to a less climate sensitive way of operating. On the other hand, a spatial relocation mainly aims to minimise the adopters' exposure to effects by seeking out a new occupation seen to be more "amenable to the continuance of their original activity" (Rickards and Howden, 2012: 243). "Contextually relevant disruptions to the status quo through the institution of novel processes and technologies are key to the idea of transformation" as Bahadur et al., (2015: 46) argue. Bahadur et al., further argue that innovative technologies and approaches enhance the possibility of transforming how vulnerable communities respond to climate variability to enhance resilience.

To enhance resilience, there is need to identify opportunities and constraints, defined by Tschakert and Dietrich (2010: 1) as "forward looking processes at the intersection of climatic uncertainty and development challenges in Africa with the overarching objective to enhance adaptation and resilient livelihood pathways". To that end, Tschakert and Dietrich advocate for what they refer to as adequate tools for anticipatory learning. For them, anticipatory learning is forward looking and a key part of reducing vulnerability/building resilience. In anticipatory learning actors:

> ought to be aware of the problem, build knowledge, diversify their ideas, reflect, communicate, develop a shared vision, and act…it involves fundamental methodological and ethical decisions such as: Managing resilience for whom? When to change from adaptation to transformation…Despite these challenges learning allows for a glimpse into how anticipation may shape resilient livelihood pathways in practice (Tschakert and Dietrich, 2010: 5).

Anticipatory learning helps smallholders adapt to variable environments (Tschakert and Dietrich, 2010). Anticipatory learning is *not* about learning subsequent to the occurrence of shocks. Tschakert and Dietrich (2010: 20) advise that "learning by shock is neither an empowering nor an ethically defensible pathway".

Addressing the concerns of this question will provide interventions that meaningfully assist smallholder farmers adapt to climate change and variability. To that end, interventions ought to address structural vulnerability.

Tschakert et al., (2014) suggest that not enough attention has been granted to addressing structural vulnerability. Instead of focusing on vulnerable groups, societal structures need to be made more equitable. Equity will have a significant impact on reducing vulnerability and will enable adaptation to be more effective (Tschakert et al., 2014). Addressing structural vulnerability is therefore a keystone of adaptation to climate.

Adaptation to climate is accompanied by two distinct challenges. The World Bank IEG (2012) says that the first challenge is adaptation to climate variability, which involves day-to-day and year-to-year variation in the weather patterns. These are experienced, for example, as chronic and extreme droughts and floods (World Bank IEG, 2012). The second challenge is adaptation to anthropogenic climate change, which involves the increase of current climate variability and the emergence of new challenges such as rise of sea level (World Bank IEG, 2012). Adapting to climate change addresses long-term climate change impacts. However, with increased, variability, impacts are changing at a rapid rate, so that one cannot rely on experience alone to plan for the future. Anticipatory learning is key to adapting to both current and longer-term change and can build adaptive capacity to deal with other stresses as well.

The concept of adaptation to climate variability overlaps partially with that of adaptation to climate change (World Bank IEG, 2012). This is evidenced by actions that merge adaptation to climate variability and adaptation to climate change, for example, strengthening extension services as this will assist smallholder farmers to address today's drought and at the same time lay groundwork for rapid response to new conditions that are emerging (World Bank IEG, 2012). Therefore, some actions that address climate variability will inadvertently contribute also to climate change adaptation (World Bank IEG, 2012). At times, adaptive strategies that address climate variability (such as extraction of ground water) may be unsustainable in future. The section below considers a typology of adaptation approaches (see Table 2).

Type	Meaning
Anticipatory, pro-active or *ex-ante* adaptation	Adaptation that takes place before impacts of climate change are observed.
Autonomous adaptation	Adaptation that does not constitute a conscious response to climatic stimuli but is triggered by ecological changes and by market or welfare changes in human systems. Also, referred to as spontaneous adaptation.
Planned adaptation	Adaptation that is the result of a deliberate policy decision, based on an awareness that conditions have changed or are about to change and that action is required to return to, maintain, or achieve a desired state.
Public adaptation	Adaptation that is initiated and implemented by governments at all levels. Public adaptation is usually directed at collective needs.
Reactive or *ex-post* adaptation	Adaptation that takes place after impacts of climate change have been observed.

Table 2: A typology of adaptation approaches (Source: IPCC, 2001)

This typology of adaptation is considered in Chapter 2 when the study explores Mr Phiri Maseko's interventions.

Burton et al., (cited in Boyd and Cornforth, 2013) distinguish between what they call "first-generation" and "second-generation" approaches to adaptation. First-generation approaches aimed at specific solutions of adaptation to specific climate change problems leading to technological fixes such as new crop varieties resistant to drought (Burton et al., cited in Boyd and Cornforth, 2013). On the other hand, second-generation approaches are context specific. They explore the context in which hazards take place and consider the nature of vulnerability "rather than merely the predicted biophysical threat" (Boyd and Cornforth, 2013: 203). Second generation approaches are closely related to transformative climate actions discussed above. Further to this, the concept of resilience has emerged as a form of adaptation (Boyd and Cornforth, 2013). Resilience takes cognisance of how to build ability to manage unexpected change (Simonsen et al., 2014). A resilience approach views the management of socio-ecological systems as critical in building the capacity of humans to adapt to climate.

However, it must be emphasised that adaptation has a long history, and since time immemorial communities have always adapted to climate variability. What can be learned from the past that can help address current impacts of climate variability?

Precolonial societies have always used innovative adaptation to climate variability. According to Adger et al.

Individual and societal adaptation to climate is nothing new, neither as an empirical reality nor as a theoretical construct. The resource irregularities offered by different climates and the precariousness which emerges from the vicissitudes of climate have both acted as significant stimuli throughout human history for social and technological innovation (Adger et al., 2009: 2).

Building on this understanding of adaptation, it is useful to draw on archaeological research of two societies in pre-colonial Zimbabwe that demonstrate how farmers built resilience as a way of adapting to climate in the past. The two examples provide evidence that societies in the past occupying different environments developed different ways of coping with adversity and constraints by innovating and adapting. Two examples are given of farmers that lived in Nyanga and the Shashi-Limpopo basin in Zimbabwe and the strategies they used to adapt to their environments. Table 3 illustrates adaptive strategies by the people who lived in ancient Nyanga and the Shashi-Limpopo basin. This study suggests that these practices may have been an influence on Mr Phiri Maseko's innovations to climate variability.

Practice	Place
Hill terracing	Nyanga
Water harvesting	Nyanga,Shashi-Limpopo
Vleifarming	Nyanga, Shashi-Limpopo
Inter-cropping	Nyanga, Shashi-Limpopo
Manuring fields using cow dung	Nyanga

Table 3: Adaptive strategies practised in the pre-colonial era by people who lived at Nyanga (AD1300-1900) and in the Shashi-Limpopo basin (AD900-present) (Adapted from (Manyanga, 2000 and Soper, 2002)

Although situated in a different context, adaptation to climate in Nyanga bares semblance to Mr Phiri Maseko's innovations to climate variability. Nyanga is in agro-ecological region 1, a high rainfall area with an average of over 1000mm per annum and is in eastern Zimbabwe. The farmers terraced hills in a bid to manage water. Soper (2002) undertook a study on how the farming communities in Nyanga in pre-colonial Zimbabwe grappled with climate variability. In the book *Nyanga* written by Soper, J.E.G Sutton writes:

Nyanga in the hills of eastern Zimbabwe is one of the exceptions, being a district where 'fossil' fields, in the form of stone terraces built on the slopes in series upon series, survive as a powerful testimony to a former farming community, and to the labour which it devoted to constructing, maintaining and hoeing these fields for the growing of sorghum and other crops (Sutton cited in Soper, 2002: vii).

Soper (2006) notes that the ruins at Nyanga are a sign of ingenuity of an agricultural society that lived from about 1300 AD to 1900. Remains of terraces (see Photograph 1) and cultivation ridges are a stark reminder of innovative water management in Nyanga. Cultivation of steep slopes involved measures to help curb soil erosion and this was expedited by piling stones one on top of the other. However, in the Nyanga case the stones formed step or bench terraces. These structures would help conserve soil and assist the process of water infiltration. Construction of these stone structures on the slopes required a lot of work.

Photograph 1: Nyanga hill terracing (Source: National Museums and Monuments of Zimbabwe, 2014)

The Shashi-Limpopo floodplains provide a contrasting climatic zone to Nyanga. The Shashi-Limpopo basin is dry with average rainfall per annum below 600mm. Archaeological studies have been carried out in the Shashi-Limpopo basin by scholars who include Manyanga (2000; 2006). Manyanga (2006) suggests that people who lived in the Shashi-Limpopo basin during periods of extreme variability beginning in the thirteenth century did not abandon the area but sought ways of adapting. They reduced vulnerability through farming vleis. The adaptation strategies included harnessing water by digging furrows in the wetlands for agricultural purposes. The Shashi-Limpopo basin constitutes an accumulation of depositional organic matter and soil nutrients (Manyanga, 2006). Manyanga also explains that even today the floodplains are still home to innovative agricultural practices by smallholder farmers. Manyanga suggests that these floodplains played a role in helping the inhabitants build resilience to rainfall variability. Farming societies in the past have always adapted to extreme weather conditions.

Humanity has built resilience to survive against adverse conditions. During extreme weather conditions, humans have learnt to be innovative. Innovation is not a new concept but attests to human ingenuity in a bid to survive dating back to pre-colonial times. The next section explores the concept of innovation.

The innovation concept

Innovation is context-specific and highly dependent on several factors, including who defines, *what constitutes* said innovation. Roling (2009: 9) views the concept of innovation as a diversity of perspectives – a real "battlefield of knowledge". This battle of knowledge has ensued over the years from the formative years of Everett Rogers (regarded in some circles as the 'father of the diffusion of innovations') as a researcher in innovation in the 1960s. Rogers' (2003) theory has been very influential in the discourse of uptake of innovations, a concept that this study interrogates in Chapter 5. However, scholars such as Reddy et al., (2010) note that before Rogers, Schumpeter in 1934 had pioneered research on the individual's role in the process of innovation (Sledzika, 2013). Reddy et al., (2010) further note that Schumpeter's conceptualisation was pivotal in the shaping of subsequent literature.

The United Nations Development Programme acknowledges that the role of individual innovators is critical in rural development (Reddy et al., 2010). Prolinnova has documented fascinating innovations by smallholder farmers in East Africa. Dobie et al., (2001) posit that in the semi-arid areas of East Africa, smallholder farmers are innovating in many creative ways. These farmers - just like my main research participant Mr Phiri Maseko - have proved to be "excellent sources of learning and simply speaking, they get their message across to their fellow farmers better than outsiders can" (Dobie et al., 2001: viii).

The concept of innovation in agriculture can be extended to include grassroots innovation. Letty and Bell (2012) suggest that grassroots innovation which smallholders have a huge say, can be a complement to conventional R&D approaches. It is generally believed that grassroots innovation has high rates of uptake as compared to linear models of technology transfer mainly because they are more applicable to the needs of smallholders (see also Letty and Bell, 2012). Those who advocate for different forms of grassroots innovation - that is, local innovation - suggest that it is a wholly locally driven process. However, this study emphasises that it would not be accurate to write in terms of a wholly locally driven process. Waters-Bayer and van Veldhuizen agree on the notion of local innovation as a dynamic process:

> Local innovation refers to the dynamics of local knowledge— the knowledge that grows within a social group, incorporating learning from own experience over generations but also knowledge gained from other sources and fully internalised within local ways of thinking and doing. Local innovation is the process through which individuals or groups discover or develop new and better ways of managing resources – building on and expanding the boundaries of their IK (Waters-Bayer and van Veldhuizen 2004: 1).

This study uses the terms local and indigenous knowledge interchangeably. The term indigenous knowledge is contested. IK has often been relegated to the backseat of research on the flawed assumption that it is unscientific (Green, 2008a). This might explain for example, the 'Green Revolution' and its emphasis on technological fixes premised on western science and its failure to acknowledge the importance of local knowledge.

In an article on South Africa's indigenous knowledge system policy, Green (2008b) grapples with the policy's dualism of indigenous knowledge and science. Green suggests:

> Research on knowledge diversity in South Africa requires focus on ontology, embodiment, and the discursive to be conducted with a steady eye on state and corporate power; a determinism to resist the dualism of 'science' v 'indigenous', and a commitment to a style of translation – and intellectual diplomacy – that can facilitate on matters epistemic and ontological (Green, 2008b: 57).

It is this study's view that I K is a fusion of both local and outside knowledge systems. The world is mobile; society is always in constant flux and therefore there is bound to be outside influences in so-called traditional practices. Achebe (1974: 46) says: "The world is like a mask dancing. If you want to see it well you do not stand in one place". In other words, society is not static. In this study, I use the definition of I K by Waters-Bayer and van Veldhuizen (2004) above.

After defining innovation, the next section explores the history of innovations in smallholder agriculture.

The 'eternal optimism' based on the 1960s benefits of western science and technology of the 'Green Revolution' (Rhoades, 1989) relegated smallholder farmer innovations to the backseat in rural development discourse. The 1960s saw the emergence of the 'Green Revolution' that was meant to be a watershed moment in the history of agricultural innovations. The 'Green Revolution' was initiated by the Rockefeller and Ford foundations of the United States of America (USA) according to the International Food Policy Research Institute (IFPRI) (2002). The two foundations established an international agricultural research system whose task was to transfer technologies to developing countries to help alleviate food shortages (IFPRI, 2002). In Africa, the 'Green Revolution' was launched by international donor organisations to try to boost food production due to perceived failure on behalf its agricultural systems. It was premised on technology transfer, known in some circles as the pipeline model, leading to mechanised farming and use of high yielding seed varieties.

The 'Green Revolution' was characterised by the linear model, wherein; "scientific research is the main driver of innovation, creating new knowledge and technology that can be transferred and adapted to different situations" (World Bank, 2006: 11). It was hailed as a landmark in the history of agricultural innovations and was packaged as *the* perpetual, on-going, solution for food insecurity. However, in countries such as Indonesia in the 1980s the 'Green Revolution' had run out of steam (Gijsbers, 2009). For example, "problems of iron and aluminium toxicity emerged in irrigated paddy fields and continuous cultivation of cereals caused serious outbreaks of plant pests such as the brown plant hopper in Indonesia" (Gijsbers 2009:109). The issues in Indonesia demonstrated the fallibility of a model, which was thought to have been replicable in all parts of the world.

Escobar (cited in Arnall et al., 2014: 99) suggests that such dominant narratives as the 'Green Revolution' have "fostered a way of conceiving of social life as a technical problem, as a matter for rational decision and management to be entrusted to the group of people – the development professionals – whose specialised knowledge allegedly qualifies them for that task". Arnall et al., (2014) weigh in by noting that the technical pathway is not wholly neutral but tips the climate change debate towards commensurability with the hegemonic narrative, neoliberalism. Drawing on these perspectives, I suggest that technical solutions have not achieved required results in rural development because they tend to 'ignore' smallholder innovations. For instance, in Africa the agricultural systems were not mechanised, had very little irrigation facilities and hence the 'Green Revolution' did not achieve much success. Dominant narratives have also been criticised for effects such as 'depoliticisation' of development. The origin of this concept is attributed to Ferguson who writes that development in Lesotho is like an "anti-politics machine" that depoliticises everything by "depriving the subjects of anti-poverty interventions of their history and politics" (Arnall et al., 2014: 99).

Recently there have been attempts at reversing the top-down models of rural development by incorporating participation of would-be beneficiaries, but such attempts have been criticised that they focus on a one-size fits all solution (Arnall et al., 2014). Pre-eminent among researchers advocating that the farmer should play a major role in the research process are Rhoades and Booth (Veteto and Crane, 2014). Rhoades and Booth (cited in Veteto and Crane, 2014: 1) say, "research should end and begin with the farmer instead of the top-down approaches." They called their concept the "farmer-back-to-farmer" model that has since been pivotal in applied research into local knowledge in agriculture and in participatory methods (cited in Veteto and Crane, 2014: 1). Auerbach (2013) adds that

Africa's food production can be increased through a range of approaches, from chemical systems characterised by high external inputs to systems premised on locally available resources and local knowledge.

In contrast, dominant narratives constitute a 'vote of no confidence' in some of the exogenous approaches meant to alleviate smallholder poverty, more so considering the interplay of stressors. The World Bank (2006) acknowledges the need for a change from the linear model to a more interactive one that is learning-based; ultimately bringing together problem recognition and generation of knowledge. However, the production of knowledge appears to be the exclusive domain of research institutions who research on behalf of rather than with the smallholders.

Research organisations have been regarded as the 'core institutions' producing knowledge and transmitters of technology to smallholder farmers (Gijsbers, 2009). Waters-Bayer (undated) corroborates Gijsbers' assertion by suggesting that the knowledge of 'experts' in research bodies and agricultural equipment companies is considered 'supreme'. However, the World Bank (cited in Vamsidhar et al., 2012: 476) suggests, "the idea that this is simply a question of better transfer of ideas from research to farmers has been largely discredited". For example, the success of the 'Green Revolution' in Mexico and India might erroneously give credence to its 'technology delivery pipeline' approach when applied in Africa. It appears that it has not met with much success given its emphasis on irrigation. The 'Green Revolution' did not achieve desired results because of its dependence on high levels of purchased inputs. The "African Green Revolution must deal with the challenges of rain-fed irrigation to be successful" (Auerbach, 2013: 22). In other words, addressing the vulnerability of smallholders is vital.

Approaches which do not require high levels of purchased inputs are more accessible to vulnerable farmers, more likely to promote resilience, and therefore also more sustainable than systems dependent on high levels of purchased inputs. Addressing vulnerability is a realisation of the limits of the 'pipeline approach' and its attendant transfer of technology (TOT) model of the 'Green

Revolution'. This led to a shift in thinking in the 1980s. Attention was shifted to what has been called the farmer first approaches (Chambers, 1989). Chambers says, farmer first approaches:

> ...empower the farmer to learn, adapt and to do better: analysis is not by outsiders – scientists, extensionists, or NGO workers – on their own but by farmers and assisted by outsiders; the primary location for R&D...is the farmers' fields and conditions; what is transferred by outsiders to farmers is not precepts but principles, not messages but methods, not a package of practices to be adopted but a basket of choices from which to select. The menu, in short, is not fixed or *table d'hôte*, but *a la carte* and the menu itself is a response to farmers' needs articulated by them (Chambers, 1989: 182-183).

In 1992, Ian Scoones and John Thomson convened a second workshop entitled "*Beyond Farmer First*". The workshop emphasised the "pluralism of different knowledges; the recognition of knowledge as not a stock but a process; seeing farmers, extensionists, scientists and others as social actors" (Chambers, 2009: xix). The year 2007 witnessed yet another conference (as already alluded to in the Introduction) at the Institute of Development Studies at the University of Sussex on farmer first approaches. The organisers gave it the name "*Farmer First revisited*". Once again, one of the major issues at stake was to make an audit of achievements from the 1987 conference. Scoones (2007) admits that a lot had changed since 1987 and smallholder farmers are even more vulnerable to new shocks and stresses such as climate change and HIV/AIDS. A pertinent question is how primary do smallholder farmers (and their knowledge) feature in the farmers first discourse and development trajectory? Numerous workshops under the banner of smallholder advancement have been held with pomp and fanfare yet poverty remains endemic among smallholder farmers. If exogenous approaches have been unable to achieve results, it is long overdue that smallholder innovations are taken seriously. It is with this in mind that the focus shifts to *good* news stories of smallholder innovations.

The importance of building smallholder livelihood resilience cannot be overemphasised. Smallholders in East Africa have been innovating as a way of building resilience. Mutunga and Critchley (2010) have documented innovations by smallholder farmers in the drier areas of East African countries such as Kenya, Tanzania and Uganda as a way of building resilience to climate variability. The smallholder technologies are like those of Mr Phiri Maseko and have shown a lot of promise. These are simple technologies mainly premised on water harvesting. Below, I paraphrase the various agricultural practices initiated by the smallholders as documented by Mutunga and Critchley (2010). A smallholder farmer in Kenya's Mwingi District named Musyoka Muindu initiated road run-off harvesting of water (see Diagram 1). Run-off water is channelled to his plot through a canal. He has been improving on this method since he started in 1993. His land is terraced. The bunds and channels he constructed serve the purpose of preserving soil moisture. Mr Muindu has assisted his neighbours in designing systems that harvest water. About 40 farmers in his vicinity have adopted his water harvesting techniques. Water harvesting has boosted his crop production thereby increasing income he realises from sale of his crops. His agricultural techniques have reduced soil erosion. This further supports my argument about how critical smallholder farmer innovations are in promoting food security in semi-arid regions.

In the Dodoma District of Tanzania, Mr Kenneth Sungula and his family utilise four hectares of sloping land for their livelihoods. He discovered his innovative adaptive strategies by accident in 1978. Mr Sungula realised that the crops he had planted in a depression were growing better than other crops close by. He started experimenting with small troughs- the *chololo pits* (see Photograph 2). The *chololo pits* harvest run-off water. They retain moisture that is vital for crops in semi-arid regions where rainfall is erratic. Over 300 farmers are said to have visited his farm in order to learn about his agricultural practices (Mutunga and Critchley, 2010). The *chololo pits* are said to be appealing for the following two reasons: (i) they boost food security, and (ii) they are easy and cheap to construct. The *chololo pits* increase the water capacity of the soil by means of harvesting and storing water and they also control erosion. These pits illustrate that innovations with simple cheap technologies are easy to replicate.

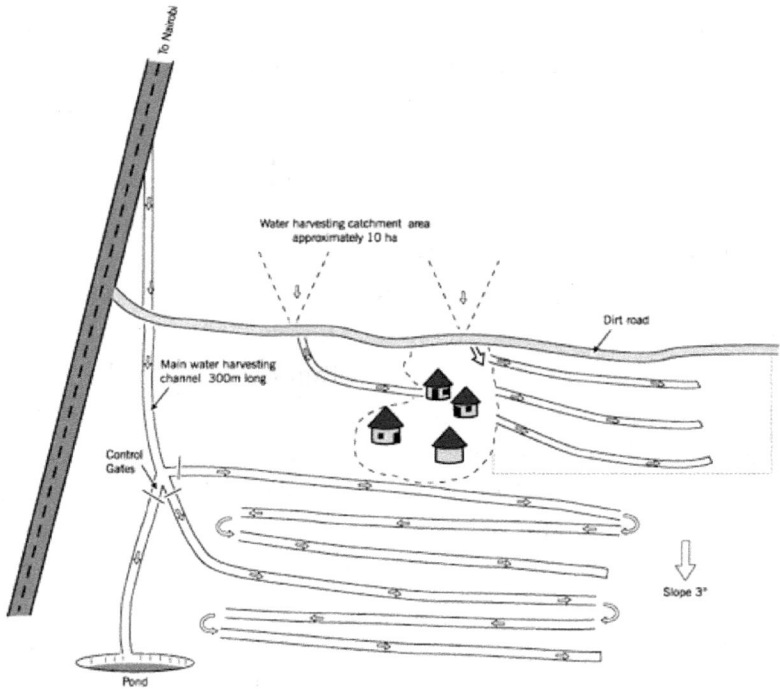

Diagram 1: Plan of Mr Musyoka Muindu's road runoff harvesting system in semi-arid rural Kenya (Source: Mutunga and Critchley 2010: 21)

The third example is that of smallholder farmer, Mr Faustino Opio, from Amuria County in Katakwi District in Uganda. Mr Opio has instituted an integrated runoff water management system. He cultivates *dambos* (swampy valley bottoms) in Katakwi. The technology he has built is self-regulatory. It allows for drainage if there is too much rain and harvests water when there is low rainfall. He divided the land into beds that were raised. These beds are separated by furrows and there is a pond below. The furrows also distribute water in the *dambo*. Mr Opio has succeeded in applying ideas he received from other farmers in East Africa to suit his own local conditions (Mutunga and Critchley, 2010). These types of innovations lead to self-sufficiency.

Photograph 2: An extension agent measures the spacing of Mr Sungula's chololo pits (Source: Mutunga and Critchley, 2010: 46)

Another recent initiative in climate-smart agriculture (CSA) is the Global Alliance for Climate Smart Agriculture (GACSA), which operates under the auspices of the Food and Agriculture Organisation (FAO). It advocates for the realisation of benefits to the people whose livelihoods are most threatened by climate change (Global Alliance for Climate Smart Agriculture Action Plan, 2014). Their strategy was launched in 2014 and is based on a multi-stakeholder approach. CSA is a multiple stakeholder network with farmers, farmer organisations and NGOs among others. The stakeholders see the need for outcomes that encourage agricultural productivity, ensure greater resilience of food systems and a reduction of GHG emissions (Global Alliance for Climate Smart Agriculture Action Plan, 2014).

Although emphasis is placed on the reduction of GHG emissions, this outcome flies in the face of the needs of smallholder farmers in rural Africa whose more pressing need is climate resilient agriculture. Of major concern is that this new network might marginalise the smallholder farmers who in most cases are left out of major decision-making processes. It appears such elitist and exclusionary approaches have low uptake among smallholder farmers. This brings the concept of diffusion of innovations into the picture. Below I explore a theoretical perspective of barriers and enablers to a food secure pathway.

Barriers and enablers to a food secure pathway

This section deploys the adoption-diffusion theory as an analysis framework of uptake of Mr Phiri Maseko's innovations that build food security, by other smallholder farmers. The adoption-diffusion theory was formulated by Rogers (Roling, 2009). Rogers (1983) says during adoption, an adopter makes full use of innovation as the best course of action to take.

Rogers (1983) explains that a decision for adoption or rejection is arrived at after the adopter considers a time-ordered sequence namely: "knowledge; persuasion; decision; implementation; and confirmation" as shown in Diagram 2. Rogers (1983) argues that in the knowledge stage, an individual hears and gathers information about the innovation. He/she asks questions such as what, how, and why. During the persuasion stage, the individual develops either a positive or a negative opinion about the innovation. The uncertainty of the innovation's attributes and the 'social reinforcement' from others influence the individual's opinions about the innovation. At the decision stage, the individual either adopts or rejects the innovation. The implementation stage means the individual puts the innovation into practice, but uncertainty may still linger in the mind of the adopter. Lastly, in the confirmation stage, discontinuance might take place in preference of an innovation or the individual might reject the innovation because he/she is unhappy with it. In

Table 11, Rogers (1983) placed adopters into five groups as follows: innovators, early adopters, early majority, late majority, and laggards.

Rogers' (1983) five categories influence would-be adopters in making a choice to take up innovations: the relative advantage of the innovation, its compatibility with potential adopters, its complexity, trialability and observability.

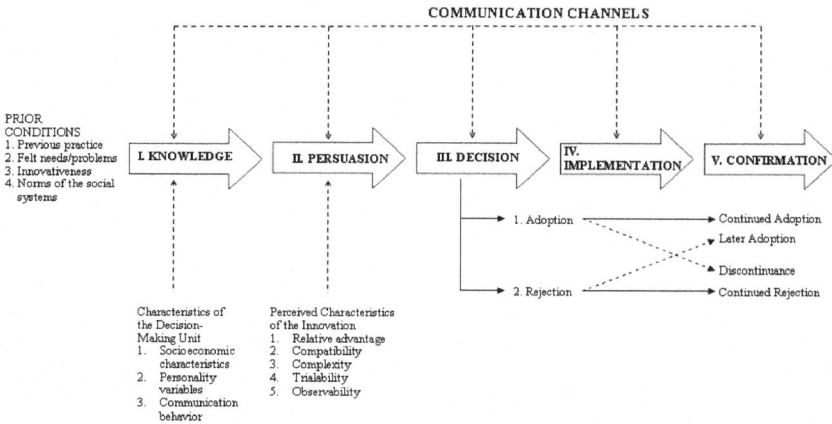

Diagram 2: A model of the five stages in the decision-making process (Source Rogers, 1983)

However, Roling (1988) critiques Rogers' (1983) theory because innovation is influenced by the farmer's capacity to learn and take risks. Roling (1988) also stresses the importance of participatory approaches characterised by collaborative learning.

The multiple-stressor approach provides a theoretical framework to probe Mr Phiri Maseko's innovative strategies to climate. Smallholder farmers in the rural areas experience both climate-related and non-climate-related stressors. Ziervogel and Taylor view stressors as:

> ...constraining factors or influences that can alter existing equilibriums and constitute a stress (drawing on the dictionary definition of stressor). Households or individuals will experience these multiple prevailing stressors as a suite of stresses (multiple stresses). In other words, a stress is something that a certain 'exposure unit' will face

31

as the result of a stressor, and so the stress experienced by different people might vary for the same stressor (Ziervogel and Taylor, 2008: 41).

Table 4 demonstrates Rogers' characterisation of adopters.

Category	Characteristics
Innovators	Willing to experience new ideas, develop new gadgets, eager to bring new innovations from outside.
Early Adopters	Role models, leaders, opportunists, economically successful, well-connected, socially respected, put stamp of approval by adopting.
Early Majority	Won't adopt without proof of benefits, innovation decision takes more time, influenced by mainstream fashion, cost sensitive and risk averse, want to hear such phrases as "user-friendly", "plug and play"," value for money".
Late Majority	Wait and see attitude, sceptics, may adopt because of peer pressure.
Laggards	Lack resources, no leadership role, interested in tangible results, hold out to the bitter end.

Table 4: Categories of adopters (Adapted from Rogers, 1983)

Non-climate-related stressors include HIV/AIDS, ecological degradation, increase in population and inter-state and intra-state conflict. These stressors inhibit the adaptive capacity to manage climate variability by contributing to increased vulnerability (Morton, 2007; Ziervogel and Taylor, 2008). HIV/AIDS, for example, reduces supply of household labour (Morton, 2007). There is dearth of literature in Zimbabwe on the impact of these stressors on how individual smallholder farmers innovate in response to climate. This study will examine these non-climate-related stressors in as far as they limit Mr Phiri Maseko's capacity to adapt to climate. An ethnographic study of this nature, one that makes a sustained observation of one smallholder farmer's vulnerability to stressors and subsequent innovative management of climate variability is limited. This study attempts to fill this void in existing literature.

Vulnerability of smallholder farmers to climate change is a result of diminishing returns from smallholder agriculture compounded by perennial poverty, absence of markets for agricultural products, intra-state conflicts and bad governance (Magadza, 2000). When addressing adaptive issues – socio-economic, demographic and policy trends which limit the capacity by communities to adapt to climate variability should be investigated (Ziervogel et al., 2006; Morton, 2007; Ziervogel and Ericksen 2010). Kabwe et al., (2009) identify several qualities that influence uptake and argue for the need to investigate the adoption process. They cite the following such characteristics: wealth e.g. ownership of cattle, lack of resources e.g. seed, lack of security of tenure and age.

Interventions should take into cognisance a "heterogeneous response" to a wide array of stressors (Ziervogel et al., 2006). A multiple-stressor environment inhibits the capacity of smallholders to adapt to climate. However, there are enablers that facilitate smallholders' adaptation to climate such as local knowledge.

Studies (Orlove et al., 2002) on locally based seasonal forecasts by farmers (ethnoclimatology) across the Andes in Peru and Palikur astronomy in Brazil respectively concluded that observing symbols such as stars, derived from local knowledge, helps indigenous people to interpret the weather. Symbolism details the way in which humans comprehend and interpret their environment. McIntosh, Tainter, and MacIntosh (cited in Crater and Nuttal, 2009: 17) assembled a volume whose authors emphasise the need to understand the human symbolic past through "social memory" or "the long term communal understanding of landscape and biocultural dynamics that preserve pertinent experience and intergenerational transmission; the source of metaphors, symbols, legends ... that crystallise social action". Crate and Nuttal (2009) rightly acknowledge that knowledge traditions are a repository of human experiences and thus an important resource for resilience and adaptation, however, I suggest that an accommodation of seasonal climate forecasts with local knowledge will help smallholder farmers build resilience to climate change.

Cultural and spiritual values influence how people remember the past and predict the future (Roncoli et al., 2009: 97). Past patterns of weather may be viewed as fond memories of one's childhood and thus serve as a "cognitive framework for remembering significant events" says (Harley cited in Roncoli et al., 2009). Memories of past droughts can be enacted through naming of children in reference to nature, and this was mainly practised during the colonial era in Zimbabwe. One such name is *Mhashu* (locusts) meaning that the individual was born during a year when swarms of locusts destroyed crops. Do these cultural beliefs act as barriers or enablers to uptake of innovations? Do cultural beliefs act as enablers that help to facilitate the creation of social capital that might help smallholders to adapt to changing climate?

The role of social capital is central to the dissemination of information. Juma (2011) has argued how informal institutions and social interactions play a major role in building interpersonal relationships and trust, which in turn act as a catalyst in knowledge, input and resource sharing. Religiosity may play a role in information sharing. There are arguments about the positive effects of religion, how it encourages "adherents to do well unto others" and arguments about the negative effect of how religiosity might cause division between the religious and those of a non-religious dispensation (Berggren and Bjornskov, 2009: 6).

Social capital is key to the process of innovating. Innovation requires knowledge from multiple sources, including knowledge users. Innovation also involves the interaction, sharing and combination of different sources of knowledge; these interactions and processes are context-specific; and each context has its own routines and traditions that reflect historical origins shaped by culture, politics, policies and power (Hall, 2010). Interactions within church groups can marginalise the already vulnerable smallholder farmers who might not belong to any church for one reason or the other. In as much as innovations may be a production of interactions, at times the innovator himself/herself is not spared from public scrutiny.

Innovation is a high stakes game for the innovator. A breakthrough may witness a high uptake of the innovation; similarly, the process of innovation may garner derisory remarks from spectators to the innovation process. Thus, there are likely to be barriers in the uptake of innovations. It is in this context that the section below addresses barriers and those conditions that might facilitate adaptation which Shackleton et al., (2013) term "enablers".

Research on barriers and enablers to adaptation is a relatively new field. Identifying barriers and enablers to adaptation to changing climate is crucial in terms of how resources are allocated and how processes will be designed to cope with them (Moser and Ekstrom, 2010). Moser and Ekstrom (2010) observe that researchers in the adaptation community have generally assumed that developed countries have lower vulnerability and a higher adaptive capacity than developing countries. However, extreme climatic events in Australia, Europe and the USA have exposed the developed countries' adaptation deficit. This has led to the emergence of a new focus on research on barriers and limits to adaptation (Moser and Ekstrom, 2010). An appreciation of barriers and enablers is critical in building resilience to shocks and stresses at the level of the smallholders (Adger et al., 2009; Shackleton et al., 2013a). Such an appreciation helps to ensure success in the adaptive process and this is summed-up by Moser and Ekstrom:

> A refined ability to identify where the most challenging barriers might lie affords the opportunity to better allocate resources and strategically design processes to overcome them. Similarly, we may learn much about adaptive capacity and ultimate adaptation success by exploring the implications of actors' skipping certain stages—and the associated barriers—in real world decision-making. Thus, the framework presented here provides a starting point for answering critical questions that can ultimately inform and benefit climate change adaptation at all levels of decision-making (Moser and Ekstrom, 2010: 6).

Barriers are "processes and conditions that hamper adaptation; the internal and external constraints that undermine or block the deployment of adaptive strategies"; on the other hand, enablers are "possible entry points for action; factors that increase the capacity of the actors to adjust practices; conditions for success" (Shackleton et al., 2013a). Shackleton et al., (2013a) suggest that hurdles (barriers) may hinder marginalised smallholder farmers from adapting to climate variability, however, smallholders have always responded to change. Diagram 3 categorises the barriers as social, biophysical, financial, psychological, political and informational. Enablers include policies, planned adaptation, opportunities and indigenous knowledge.

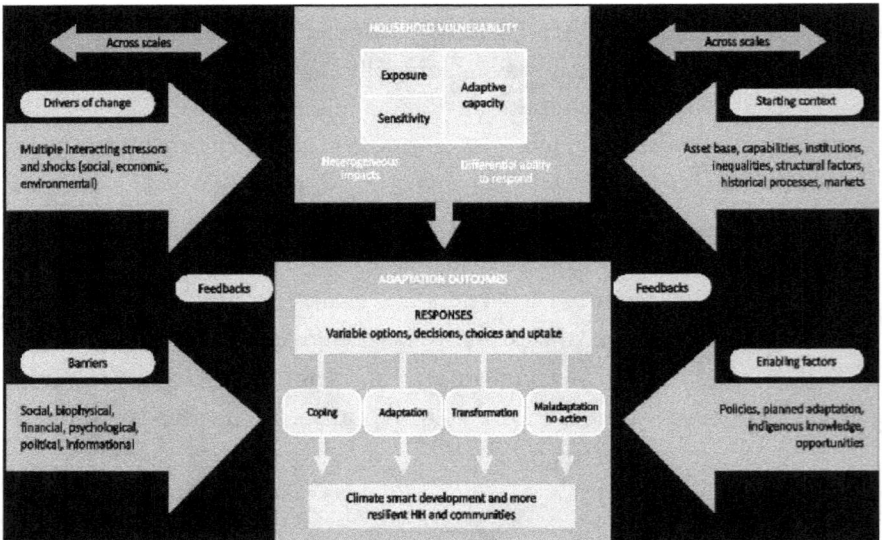

Diagram 3: Conceptualising barriers and enablers (Source: Shackleton et al., 2013)

Based on the above insights, I use Diagram 3 on the conceptualisation of barriers and enablers to analyse the hurdles to a food secure pathway by smallholder farmers in Zvishavane. In the introduction, a key guiding principle was to identify why smallholders are still 'there'. Identifying barriers, helps address this quandary. The question examines why efforts to help smallholder farmers adapt to climate have not been successful. The barriers and limits framework was used to assess why Mr Phiri Maseko 'rhymes' with his audience, an issue I address in Chapter 4. I define the concept of rhyming with an audience to demonstrate why his innovations are accepted by other smallholder farmers.

Shackleton et al., (2013b) demonstrate that vulnerability leads to different responses. These outcomes might help build resilience and lead to adaptation or coping. At times responses to vulnerability lead to maladaptation thereby increasing food insecurity.

Conclusion

After arguing for the need of the urgency of now - the need to re-evaluate our values in a changing climatic environment, this study shifts its gaze from prophecy of old, "Thou shalt perish" to what the study calls the prophecy of today, "Thou shalt not only survive but thrive…if". Central to the following chapter is to showcase the efficacy and cost-effectiveness of smallholder innovations by drawing on the innovations of Mr Phiri Maseko. In other words, this book argues that to fully grasp *how* he built resilience, it is imperative to explore where he is coming *from*. Adger et al., (2009) explore how individual, social abilities and ingenuity that enable successful adaptation may be harnessed. In other words, documenting how Mr Phiri Maseko began innovating is vital in addressing how he navigated barriers and enablers in building resilience to climate variability. Chapter 3 attempts to establish his motivations, the barriers he encountered and how it 'all began'. Through appraising the enablers that helped to facilitate his marriage of water and soil the chapter provides intimate insight into the innovative chronology of a smallholder farmer and lessons for future farming adaptation in similar arid conditions. Therefore, a different approach, that offers hope, that counters a doomsday approach, is the way forward if marginalised societies are to be assisted. The study, therefore, offers hope by presenting what Tepperman (2016) terms the "often-overlooked *good* news stories". One such story is that of Mr Phiri Maseko, a *good* newsmaker.

Chapter 2

Thou shalt not only survive but thrive, if…: *Hurudza* and mediator

Work hard and fear nothing - Rev Ndabaningi Sithole (Mr Phiri Maseko's former teacher).

I've been thinking of petitioning to change the spelling of adversity to *add*versity. Frankly, the value of adversity has an amazing way of adding things to our character and our attitude that nothing else can, as long as we are willing to consider the value therein – Del Sesto (2013)

Introduction

This chapter is a prelude to Mr Phiri Maseko's innovations for managing climate variability. The chapter demonstrates that if water and soil are in holy matrimony, smallholders will not only survive but also thrive in the wake of increased climate variability. Ethnographic research helps to present Mr Phiri Maseko's history of marrying water and soil and thus, yields insights into the factors that influenced his success at taming an unproductive dry piece of land into (in his own words) a "water plantation". The road he travelled offers valuable lessons for other smallholder farmers in situations similar to his, as well as the need to create interdependencies as encapsulated by his marriage of water and soil. An instinct for mediation lies at the heart of Mr Phiri Maseko's innovations to climate variability and this has been instrumental in moulding him into an opinion leader. Drawing on the idea of Mr Phiri Maseko as an officiant of a marriage between water and soil is central to his innovation. This 'marriage' denotes an element of conviviality. Nyamnjoh (2002: 111) asserts, "broadly, conviviality involves different or competing agentive forces which need a negotiated understanding". Conviviality is about reconciling differences. Thus, the essence of the evolution of his innovative system appears to be premised on the recognition that survival in semi-arid Zvishavane is achieved when one 'marries' Shona agricultural practices and modern technologies. Thus, this study, in the tradition of optimist Buckminster Fuller, has faith in the role of technology. However, deployment of technology in the rural areas should take on board smallholder innovations. Surviving and thriving in an environment characterised by adversity is about hybrid interventions, Mr Phiri Maseko-esque.

Maya Angelou's poem, "*Still I rise*" aptly captures Mr Phiri Maseko's determination to succeed against all odds. The poem's theme is resilience and triumph over adversity. Therefore, I argue that the adverse effects of climate variability and years of harassment by the colonial authorities strengthened Mr Phiri Maseko's resolve to build resilience and attain the status of a *hurudza* (an accomplished farmer) in rural Zvishavane. A *hurudza* was the lifeblood of the Shona economy. The term *hurudza* has evolved to reflect contemporary challenges and adaptability to climate fluctuations. In direct contrast to the *hurudza* is a *simbe* (the lazy farmer). *Simbe* were the subject of ridicule from village jesters. Their opposite, the *hurudza* served as a source of inspiration and this is the story of Mr Phiri Maseko presented in this study. While I am aware of the danger of portraying him as a paragon of virtue fighting hunger in semi-arid southern Zimbabwe – what Piot (1999: 25) terms an "overly celebratory, romanticised image" I contend that there is also a downside to his story. However, like Piot's encounter with the Kabre in Togo, my experience with Mr Phiri Maseko was also extremely positive and this shapes my favourable opinion of him. This chapter chronicles the history of Mr Phiri Maseko. Interrogating his history is important in order to try to account for the evolution of his innovative agricultural practices. I explore Mr Phiri Maseko history from his days in school, his involvement in political activism, his innovations in the post-independent era and the role of religion in his life. I portray him as a smallholder who encapsulates the concept of *hurudza* and that redefines the concept as an 'enviro-preneur' a term that is explained in this chapter.

"Days are numbered"

"Days are numbered," lamented Mr Phiri Maseko, "I am down to one eye and one ear" (see also Lancaster, 2008) as he sat on the couch at his home in rural Zvishavane where his legend took flight. This is about the reality that one of his ears and one of his eyes were no longer working because of torture at the hands of security agents in the 1970s during Zimbabwe's liberation struggle. He was visibly elderly - he walked with a limp and hardly visited his water plantation except for a few occasions when he had visitors. He used the expression of numbered days very often as he reminisced about his golden age. Even as he aged by the day, his humility and penchant for forgiveness (despite the disability he suffered at the hands of colonial authorities) made him a revered elder statesman among smallholder farmers in southern Zimbabwe.

Fences at his plot are broken, water canals are filling with leaf litter and thus the 'Eden Project' appears to be in decline. The plot is now a shadow of its former self. Has his 'steamroller project run out of steam?' As he relaxed on a sofa at home, he was consciously aware that his best days as an innovator were behind him, thus, he was no longer the 'Great Helmsman' he used to be, when he steered his project to a cult status. When confronted with the reality of how his pet project appeared to be in decline, he was quick to remind his audience that *"Zvinoda akazviona"* (it is only those who witnessed his arduous journey of innovating who would appreciate the hard work that went into this project). *"Hazviparari izvi"* (the project will not collapse) he boldly declared. To a casual visitor, the present state of his plot might betray its eternal significance to smallholder livelihoods. Yet the apparent neglect of the project due to his old age did not in any way diminish the ripple effects of its success among other smallholders. It has proved to be resilient to climate variability in the past and will continue to do so in years to come. This ecological journey made him famous. At the time of his death, his charisma was still burning at peak power. No doubt, his legacy will live on through the continued viability of the project.

The *hurudza* concept

This study demonstrates that the journey of Mr Phiri Maseko's innovations cannot be contained in a single story. As Adichie points out:

> I have always felt that it is impossible to engage properly with a place or a person without engaging with all the stories of that place and that person. The consequence of the single story is this: It robs people of dignity…Stories matter. Many stories matter…when we reject the single story, when we realise that there is never a single story about any place, we regain a kind of paradise (Adichie, 2009: 5-6).

Adichie (2009) says that a single story is inaccurate and oversimplified. This chapter suggests that an analysis of the story of Mr Phiri Maseko can be realised by telling many stories behind his story of "marrying water and soil".

As Mr Phiri Maseko walked me through his eventful 85 years of life while we were seated in his living room during the month of October (2012), I realised that there are many stories about him. Stitching these stories together is critical in understanding Mr Phiri Maseko, a person of Malawian origins. In the neighbouring Shurugwi district for example, the locals refer to homes owned by people of Malawian origins as *pamuchawa* (a person of Malawian origins). The term may be considered prejudicial, but on closer examination it may stem from jealousy as *mabwidi* (derogatory term to reference people of Malawian origins) are successful. Even in contemporary drama, people of Malawian origins are subject to ridicule. Due to inter marriages between the so-called *mabwidi* and the local Zimbabwean people, there has been a lot of interaction leading to the dilution of some customs of the people who originated from Malawi. This is typical of what Kopytoff (1987) terms the "African frontier". Kopytoff suggests that the African frontier can constitute an arena for "cultural-historical continuity and conservation" over and above being a space for transformation. However, drawing on my ethnographic fieldwork, the study suggests that Mr Phiri Maseko's innovations to climate variability are a fusion of both "cultural-historical continuity" and his transformation in Zvishavane where he lived among mostly the Shona. His perception of soil and water is rooted in Shona cosmologies and it may be that conviviality that enabled him to become a reputable *hurudza*. The story of him as *hurudza* is one of relentless experimentation in a bid to survive in semi-arid Zvishavane. It is a journey of hard work and fearlessness and a strong belief that he could conjure magic and tame a piece of land in semi-arid Zvishavane into a productive plot – a journey that can be traced from his childhood.

Mr Phiri Maseko was born on 2 February 1927 in Zvishavane. His father Amon, known for his generosity, migrated to Southern Rhodesia (now Zimbabwe) from Malawi. He came to Zimbabwe seeking employment in around 1914. He initially worked at Inyathi Mission near Bulawayo (Zimbabwe's second largest city, 430 km southwest of Harare) where he met and soon married his wife Thandi. They moved to the then Shabani (now Zvishavane) and worked at a local bakery. Mr Amon Phiri was said to be a talented singer and soon attracted the attention of prominent worshippers of the Church of Christ. One such leader of the Church of Christ was Mr Hlamvelo who taught at a mission school named Dadaya about 15 km to the west of Zvishavane. Mr Hlamvelo immediately recruited Mr Amon Phiri to teach among other schools, Mbilashaba and Msipane that fell under Dadaya Mission for administrative purposes. This created opportunities for his children to attend school. Mr Amon Phiri was allocated a plot in rural Zvishavane to supplement his salary. He got the plot through the assistance of Sir Garfield Todd (a story that will be dealt with more extensively elsewhere in this chapter). He proved to be an enterprising farmer as he sold most of the produce to local people and soon owned many cattle. At some stage, he owned as many as 35 cattle, according to his son. Mr Amon Phiri baked bread in anthills and redistributed the bread free to the local people. He was a *hurudza* and made a huge impact in the life of the young son Mr Phiri Maseko. However, what is a *hurudza*? Below I define the term *hurudza* in more detail.

A *hurudza* is an enterprising farmer. *Hurudza* occupied privileged spaces in Shona societies. They were vital cogs in the Shona economy. Mawire defines *hurudza:*

> ...both in the singular and plural, means an agricultural baron or barons... individuals who had proved to be hard-working, productive farmers. They were thus accorded a social standing comparable to that of a highly successful industrial tycoon in the capitalist societies of today. He was well connected socially and accordingly played an influential role in all Shona tribal and national affairs (Mawire, 2013).

The meaning of *hurudza* has changed over the years from the pre-colonial to the present era. Scoones (1988) explains that the present definition of *hurudza* encompasses entrepreneurship, and the *hurudza* created an aura of invincibility. The *hurudza* could also be equated to the present day 'A List' celebrities. The ordinary people were in awe of the *hurudza's* achievements. Even today, when revellers are drinking beer, they occasionally quip, "*Tohudzasa isu vana vehurudza*" (We are enjoying ourselves children of *hurudza*). The implication of this remark is that the term *hurudza* is synonymous with hard work and prosperity. The revellers would be trying to convey the message that they have worked hard (just like a *hurudza*) to raise money to make merry. They could also be trying to say that they can afford to be happy because their *hurudza* parents work hard and raise money to fend for them. The *hurudza* is still regarded highly.

In the past, a *hurudza* formed part of the elite in Shona societies together with the *hombarume* (accomplished hunter/s) and rain petitioners. The *hurudza* owned a lot of livestock that provided milk, meat, hides and for use as draught power. A *hurudza's raison d'être* was to help enhance food security in Shona societies. The *hurudza* planned well ahead of the agricultural season by using herbs to preserve seed and keep their *chipani* (cattle used as draught power) well fed. The *hurudza* preserved crop residues as stock feed for lean times. They produced surplus food, which they stored. In the event of famine, families who had no food security engaged in *kuzvarira* (marrying their daughters to the *hurudza* in exchange for grain). Most of the *hurudza* practised polygynous marriages and raised many children. The motive behind such polygynous marriages is contested. In his analysis of polygynous marriages among the Kabre of northern Togo, Piot notes that:

> It appeared from the literature on Kabre, for instance, that male elders came to sit atop various hierarchies by controlling the labour of others of women and children through the practice of polygynous marriage (which added labourers to the homestead's workforce, and thus might benefit those males who were in charge) and of younger men through a system of work groups (which was run by male elders and, it appeared might be exploited to their advantage)…what I found, however, was quite different. Men who married polygynously had to work harder not less, since Kabre women do not farm. Thus, a man with two wives (and the additional children adding a second wife brings) had to work twice as hard, since he was the sole provider of cultivated food for the family (Piot, 1999: 8 – 9).

The analysis here agrees with Piot to the extent that the man worked very hard to ensure that the family's livelihoods were secured. However, I argue that in Shona communities because the polygynous *hurudza* was an industrious individual, this inspired his wives to follow suit. The wives (ably assisted by their offspring) compete through hard work in order to gain the favours of the husband. In a way, the husband empowered his wives who would also become *hurudza* in the event of his demise. This ensured that the family of the *hurudza,* who had passed on, did not starve but continued to produce adequate grain for their consumption and even for the strategic communal grain reserve during times of scarcity. The concept of *hurudza* was not only confined to men. Hardworking women could also become *hurudza*. The *hurudza* status was earned and had to be safeguarded at all costs and this meant that the *hurudza* had to be a relentless experimenter so as to keep producing a lot of grain. Experimentation is at the heart of Mr Phiri Maseko as a *hurudza*.

He told me an interesting story that demonstrated his *uhurudza* and desire to experiment:

When I was at Dadaya Mission in the 1940s, Sir Garfield Todd gave me a mango seed to plant. Back then, it was believed that if you planted a mango seedling, the very year the tree bore fruit, you would die. When I brought the mango seed home, my elder brother warned me that I was playing with fire. I told my brother that nothing was going to stop me from planting the seed. I then planted the mango seed and when eventually it bore fruit Zephaniah did not die. Here I am still enjoying fruit from the mango tree (showing me the mango tree). Aah Zephaniah *bakithi*!

Hurudza loaned their livestock (*kuronzera*) to those who did not have livestock. This cushioned the have-nots who would use the loaned livestock as sources of milk, thus supplementing their diet. The recipients of the loaned cattle could also use the livestock as draught power ensuring that they would harvest enough to feed their families. Philanthropy was an important attribute of the *hurudza*.

Another attribute of the *hurudza* was the ability to adapt to one's environment. The *hurudza* studied rainfall patterns in their environments so as to make informed decisions of crops to grow. According to Mr Phiri Maseko, *hurudza* in the semi-arid Zvishavane area primarily grew small grains such as millet because they are drought resistant. The *hurudza* kept livestock that was suitable to agro-ecological regions where they lived. For example, in dry regions such as Zvishavane, livestock kept by farmers are mainly cattle, donkeys and goats. Donkeys and goats are drought resistant livestock. Most of the farmers in Zvishavane keep donkeys for use as draught power. *Hurudza* such as Mr Phiri Maseko rear domesticated guinea fowls because they are tolerant to dry regions and can survive in variable conditions.

Yet the *hurudza* had a nemesis in the form of the *simbe*. The *simbe* was the quintessential *persona non grata* of Shona societies and inevitably, the laughing stock of the village. Mawire elaborates about the *simbe*:

> as a result of the stigma attached to this social label and the psychological stress it produced in the unfortunate victims of this tribal

intolerance, very few people remained *simbe*...it proved to be an extremely effective way of deterring people from developing lazy habits and of ensuring a high standard of living for everybody in the country (Mawire, 2013).

The view above concurs with my findings that the *simbe* label worked to deter would-be *simbe*. *Hurudza* such as Mr Phiri Maseko helped to reduce the incidence of *husimbe* (laziness) in rural Zvishavane.

Families would not want their daughters to be married to *simbe*. This inspired individuals to be *hurudza*. However, if a prospective son-in-law was poor and could not afford to pay bride wealth, he would engage in *kutema ugariri* (working for in-laws in lieu of paying bride wealth). *Kutema ugariri* also served the purpose of ensuring that daughters were married to hard working sons-in-law. The modern day *kutema ugariri* in a present day market economy can be equated for example, to some of Mr Phiri Maseko's adopters who work for him in return for technical advice. Two of Mr Phiri Maseko's adopters visited him to do different chores at his plot as a way of *kutema ugariri* (see Photograph 1). For example, the adopters assisted Mr Phiri Maseko in digging canals on his plot. In a way, they acted as his apprentices. Over the years, Mr Phiri Maseko helped them to hone their agricultural skills during that apprenticeship period and today they have become *hurudza*. The concept of *kutema ugariri* in a way was therefore a rite of passage for Mr Phiri Maseko's adopters *en route* to becoming *hurudza*.

Mr Phiri Maseko further demonstrated his *hurudza* skills as the senior community supervisor for the Lutheran World Federation Water Programme in Zvishavane and Mberengwa districts between 1981 and 1986, where he gained invaluable experience in working with other smallholder farmers. During his stint with the Lutheran World Federation, Mr Phiri Maseko was mainly helping smallholders to construct shallow wells and develop water harvesting techniques ideal for a semi-arid region. He encouraged other smallholder farmers to produce not only for their families but the whole community as well.

Photograph 1: Two of Mr Phiri Maseko's adopters working at his plot in return for technical advice

From the interviews, I conducted, the concept of *uhurudza* encapsulated the idea of a strategic grain reserve. Society would count on *hurudza* for purposes of *kushuzha* (procuring grain in times of scarcity). *Kushuzha* could involve Shona societies further afield who would then travel hundreds of kilometres to secure grain. The benevolence of the *hurudza* aristocracy, to use a phrase by Mawire (2013) was therefore not only confined to their locality. Benevolence and the ability to be innovative, make the right decision of what to plant, knowledge of when to plant, boost soil fertility and preserve grain is what makes a *hurudza* a *hurudza*. After interrogating what a *hurudza* is, the section below elaborates the making of Mr Phiri Maseko into a latter day *hurudza*.

The making of a latter-day *hurudza*

I begin by elaborating on the making of a traditional *hurudza* in proverbs/idioms (see Diagram 1) and folk songs. This resonates with the process of shaping Mr Phiri Maseko as a productive farmer. The Shona proverb *ane maoko maviri haatsvi nenyemba* (He who has both hands will not be burnt by hot beans) means that you must be creative in ensuring that your hands will not be burnt by the hot beans. Transferring beans from one hand to the other as the other recovers from the heat until the beans are no longer hot prevents one's hands from being burnt. Mr Phiri Maseko had been allocated a piece of land next to his father's plot by the village head (after consulting the chief) is the custom in rural Zimbabwe. After his dismissal from his job at the RR, he turned to farming and started tilling his piece of land. This is the time he realised the importance of perseverance.

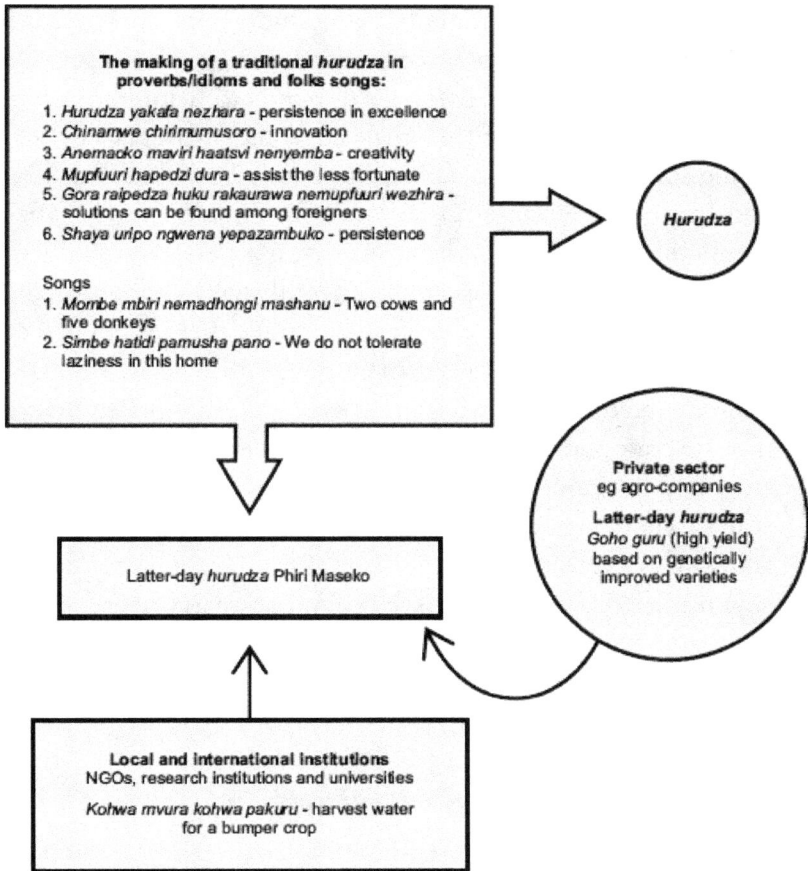

Diagram 1: Shaping of hurudza in temporal terms

The other proverb *shaya uripo ngwena yepazambuko* (You must persevere like a crocodile that waylays its prey at crossing points) encourages people to be patient and determined in whatever they do. The crocodile must be very patient until its potential prey crosses the river and only then, it attacks. A smallholder farmer in neighbouring Chivi district, a follower of Mr Phiri Maseko owns a field comprised chiefly of a type of soil known locally as *chinamwe* (heavy clay soils with poor drainage). When asked how she creates high yields, she said, *"Chinamwe chiri mumusoro"* (Clay soils do not exist in the field but in the head as a psychological problem). The proverb used by this smallholder farmer continued to resonate in my mind as I was going back to my 'base' at the Phiri Maseko homestead. The *chinamwe chiri mumusoro* smallholder farmer has continuously worked to add fertility to her soil by adding compost manure. She has also instituted water drainage by means of digging small ponds to help drain her plot. In some places, she excavated the clay soils and replaced them with fertile soil. This smallholder farmer is now a well-acclaimed *hurudza* and has won awards in several smallholder farmer field day competitions. The proverb means that if a person works hard he or she can overcome any adversity. These proverbs encouraged *hurudza* to be patient and persevere (like a crocodile that resides at a crossing point) in order to achieve their goals. Mr Phiri Maseko amply demonstrated this. He persevered over the years as demonstrated by how he continuously innovated to improve his agricultural practices.

Hurudza were also inspired to work hard through Shona folk songs. The songs advocate zero tolerance to laziness. One of the songs relevant to this book is *Karingahore* (He/She who stares at the sky) recorded by Phillip Svosve (a Zimbabwean singer):

Karingahore
Simbe hatidi nyope hatidi pamusha pano
Nyope hatidi simbe hatidi pamusha pano
Tora mapadza muroora utevere kumunda
Tora foshoro mukwasha utevere nengoro

Chorus

Gogogogo Karingahore kumadokera hehahe mvura yaturuka
Gogogogo Karingahore kumadokera hehahe mvura yaturuka
Ndimwi here makaramba kurima kumadokera hehahe mvura yaturuka
Ndimwi here makaramba kurima kumadokera hehahe mvura yaturuka

He/She who stares at the sky

We do not tolerate laziness at this home
We do not tolerate laziness at this home
Take a hoe our daughter-in-law and follow us to the fields
Take a shovel son-in-law and follow us to the fields with an ox-drawn scotch cart

Chorus

Rain has started falling from the west
Rain has started falling from the west
Is it you who refused to farm, rain has started falling from the west
Is it you who refused to farm, rain has started falling from the west

Most of these songs are presently sung at field days in rural Zimbabwe where smallholder farmers gather to display their agricultural produce as a way to inspire others. Some of these songs have been recorded by artists and reach a wider audience through airplay and record sales. Songs in Shona culture play an influential role in dissuading people to be lazy but to work hard as demonstrated by the *hurudza*.

The *hurudza*'s innovation takes time to develop (as already alluded to) and can be traced to the individual's childhood. I suggest that one's childhood has a seminal influence on the creativity that one might unleash in later life. Childhood experiences influence innovation as illustrated by the story of a girl I came across while I was doing my fieldwork in 2012. What struck me most about her was the 'crown' made of leaves that she was wearing (see Photograph 2). I gave the young girl the title, 'African Queen'. When I was back at the Phiri Maseko homestead in the evening I began to reflect on the picture of the young 'African Queen'.

51

Photograph 2: The young 'African Queen' wearing her 'crown'

Her picture told me many stories, and I was struck by her amazing level of creativity. For me seeing her wearing the lovely 'crown' was a fortunate stroke of serendipity. Adversity nurtures innovation. Poverty, hunger, disease, and drought in rural Zimbabwe create an environment for creativity for children who eventually become innovators. Adversity therefore infers that such children are hardwired to be innovative. They learn to survive in the face of inadequate provisions from their parents. The crown of the young 'African Queen' conveyed the message that in the absence of a hair salon, she is compelled to be creative and use resources in her environment as a survival tactic. Creativity such as this is prevalent among children as they try to come to grips with the reality of their vulnerability caused by rainfall variability and many other stressors that are characteristic of semi-arid rural Zimbabwe. The countryside is replete with many of these young African 'queens' and 'kings' whose creativity is astounding, and for me this demonstrates that innovation can be innate. Again, marginalisation motivates individuals to be innovative.

Marginalisation of migrants of Malawian origins in Zimbabwe spurred some individuals to be innovative. This view is echoed in separate interviews at the University of Zimbabwe I held with Prof George Kahari (Department of African Languages and Literature) and Mr Anusa Daimon (History Department) (12 July 2012), it emerged that people of Malawian origins are said to have a long genealogical history as risk takers and trying new things. It was unimaginable that as early as the 1920s people from as far as Malawi could migrate across harsh terrain to a foreign country on foot. They had to contend with crossing crocodile-infested rivers, diseases and dangerous wild animals. Thus, they had to be innovative to survive. The people of Malawian origins are said to be innovative pioneers in for example, welfare societies in Zimbabwe (Kahari, 2012). The fact that they were far from home meant they had to help each other. Mr Phiri Maseko's innovations can be seen through such lens.

The genesis of Mr Phiri Maseko's innovation could therefore be traced to his childhood. Having grown up in rural Zimbabwe, he came face to face with adverse effects of rainfall variability worsened by a multiple-stressor environment. At an early age, he could have observed his parents toiling to put bread on the family's table in dryland Zvishavane. Could he have made a conscious decision then that survival in semi-arid Zvishavane was about harnessing water? His childhood no doubt is characterised by creativity, according to testimonies from his contemporaries Messrs S Sibanda (village head), Dinga, Magaya and F Sibanda. Stories abound about Mr Phiri Maseko's childhood. All his contemporaries concurred that he was never afraid to experiment even at a very early age. The stage marks the genesis of his journey to the water harvester's world.

I also discuss the views of some of his neighbours that are also discussed at length in Chapter 4. The water harvesting skills of Mr Phiri Maseko have attracted many schools of thought especially among smallholder farmers in rural Zvishavane. Some of his neighbours allege that he has "magical powers" as evidenced by how he attracts people who visit him, including *varungu* (whites) who "assist him financially". Some attribute his success as a farmer to the use of *divisi* (a crop yield enhancing charm). Some hail him as a hard worker who has inspired them to reach new heights of food security. The village head of Hlupo, a contemporary of Mr Phiri Maseko, says:

I first met Mr Phiri Maseko when we were young as small boys herding cattle. There was always something unique about the way he did his things. For instance, the way he combed his hair was different from everyone. Even if you take a closer look at his hair today, you find that there is something unique about the way he combs it. I remember he used to ride a donkey to school. The way he decorated the donkey was different from everyone else. Everyone at school knew Mr Phiri Maseko's donkey by its decorations. He was different from his brothers and in the Msipane and Siboza areas there were two outstanding young men – Mr Phiri Maseko and (pauses)… I have forgotten the other one. What he is doing now does not surprise me (Interview with Mr I Sibanda, 24 October 2012).

On the other hand, Mr Shingayi, a critic of Mr Phiri Maseko, vowed that he will never "waste time" following Mr Phiri Maseko's methods. He spoke plainly and showed me his granary, which was full of grain.

Mr Shingayi took me through what he called a "Sermon on the Mount" about his *uhurudza* status he achieved using agricultural practices he learnt at college. He displayed his agricultural certificates to try to prove his point. He waxed lyrical about his achievements, oblivious of the possibility that some of his methods of conserving soil moisture might be traced to Mr Phiri Maseko. One adopter I interviewed was motivated by Mr Phiri Maseko's agricultural practices.

Mr Magogomere had this to say:

The first time I saw Mr Phiri Maseko is when he came to our school Utongani High (in Ward 6) to talk to us about water harvesting. A few years before that, his organisation, the Zvishavane Water Project, had constructed water tanks at Msipane Primary School just a few metres from our classrooms. He took us to the tanks and began talking to us about the importance of water harvesting. He taught us how the water tanks at Msipane harvested water and that the water would be used for market gardening and cleaning our toilets. I was impressed by the water harvesting methods espoused by Mr Phiri Maseko. After finishing school, I could not get a job and decided to approach Mr Phiri Maseko to teach me more about water harvesting since I had been allocated a plot. However, he said I should assist him in doing various chores at his plot instead of paying him a fee. During the days I spent as Mr Phiri Maseko's apprentice, I interacted with his neighbours. Some alleged that Mr Phiri Maseko used *divisi* to enhance his yields. Said one of them: "What he does is not normal, how can he work like he is possessed. I do not believe this; he is assisted by *zvitokoloshi* (goblins)...I will have difficulties in copying Mr Phiri Maseko's methods..." These allegations did not deter me from learning Mr Phiri Maseko's agricultural methods. He impressed me especially with the methods he uses to drain his plot when there is too much water on his plot. Look I also planted sugar cane and bananas like what he did. These plants take a lot of water so that they help in draining my plot (Interview with Mr Magogomere, 7 December 2012).

Such views portray Mr Phiri Maseko as an enigma. My ethnographic fieldwork portrays him as an unerringly forthright, jovial, modest, open-minded and easily accessible smallholder farmer whose unique agricultural methods are a product of his innovative mind, nurtured in the context of the hazards and opportunities in semi-arid southern Zimbabwe.

There are a host of people who played an important role in the life of Mr Phiri Maseko. His father, Mr Amon Phiri (a Malawian immigrant) taught his son to accept who he was, to appreciate that he was of Malawian descent. Baldwin (cited in Rankine) advises that accepting who we are is the first step towards being a winner in the face of adversity. He writes, "Our humanity is our burden, our life; we need not battle for it; we need only to do what is infinitely more difficult – that is, accept it" (Baldwin cited in Rankine, 2015).

Rankine (2015) adds that to accept the self is a way of discarding the racist gaze. My data proposes that it was out of the realisation that being a person of Malawian descent should not hinder him to experiment on innovations that would help him to adapt to changing climate.

His father motivated him to appreciate the importance of agriculture in ensuring food security. The methods of farming Mr Amon Phiri implemented were imprinted on the mind of the young Zephaniah as ducklings are to the sound of their mother's call: innate and hardwired. Mr Amon Phiri had learned many things form his association with Sir Garfield Todd. In addition to the influence of his father, the story of the making of Mr Phiri Maseko as a latter-day *hurudza* would be incomplete without identifying the role of local and international institutions, NGOs and agro-business companies. I will identify several individuals at both local and international institutions from his days in school.

Mr Phiri Maseko started his education in the 1930s. He finished his Sub A and B (Grades 1 and 2) and Standard 1 at Sivanga Primary School in Zvishavane. In 1941, he moved to Dadaya Mission where he did his Standard 2 to 6. It was during this time that he realised the importance of growing fruit trees because he harvested much fruits at Dadaya Mission. The principal of Dadaya Mission, Sir Garfield Todd, a missionary from New Zealand, gave Mr Phiri Maseko a mango seedling to grow at his home (this story is returned to later in the chapter).

Sir Garfield Todd was the Prime Minister of colonial Southern Rhodesia (Zimbabwe) from 1953 to 1958. He had come to the then Southern Rhodesia as a missionary of the Church of Christ in the 1930s. Sir Garfield Todd replaced his compatriot, a Mr Bowen who had established Dadaya Mission (15 km to the west of Zvishavane) at a site, which is now known as Old Dadaya. Dadaya Mission then moved to its present site near the Ngezi River in the late 1940s courtesy of Sir Garfield Todd who donated part of his Hoknoi Ranch to the school. Sir Garfield Todd taught at Dadaya Mission with his wife Grace. The Todds played a pivotal role in advancing the cause of African education. The Todds "wrote the whole education curricula from Sub A up to Standard 6 plus two years of teacher training. Their education schemes became the foundation of the education of all black people" (Mhlanga, *The Daily News* 20 January 2014). The school motto was "Education for life" and this resonated with Sir Garfield Todd's policy. The school curricula included such subjects as Agriculture and Building. In an interview, I held (on 24 January 2014 in Gweru) with Dr C Msipa the former Governor of the Midlands Province in Zimbabwe, I learnt that it was Sir Garfield Todd who pioneered the planting of crops in lines, an innovation that was greatly resisted at the time. This innovation however, eventually gained wide acceptance because it proved to be productive. Sir Garfield Todd was a critic of white minority rule and fought for the civil rights of blacks in the then Southern Rhodesia. He was to support the freedom fighters in the 1970s and was imprisoned by the Ian Smith government in the late 1970s. His political views inspired his students such as Mr Phiri Maseko to become involved in politics in order to liberate Zimbabwe. In an obituary dedicated to Sir Garfield Todd, Keatley and Meldrum (*The Guardian*, 14 October 2002) write that Sir Garfield Todd's philosophy derived from the Bible: "Just keep throwing your bread upon the waters; if you're lucky, it will come back as ham sandwiches". Sir Garfield Todd's philosophy helps to explain Mr Phiri Maseko's relentless experimentation in a bid to manage climate variability.

As an experimenter, Mr Phiri Maseko confessed that at times he would end in failure. However, the creator of Harry Potter, Rowland says that failure has fringe benefits. She says failure gave her inner security:

> Failure taught me things about myself that I could have learned no other way. I discovered that I had a strong will…The knowledge that you have emerged wiser and stronger from setbacks means that you are, ever after, secure in your ability to survive. You will never truly know yourself, or the strength of your relationships, until you have been tested by adversity. Such knowledge is a true gift, for all that is painfully won… (Rowland, 2008).

Adversity made Mr Phiri Maseko work hard. At Dadaya Mission he first learned what was to become his maxim – "work hard, and fear nothing" from one of his teachers, the Reverend Ndabaningi Sithole. Rev Sithole was one of the founding fathers of the Zimbabwe African National Union Patriotic Front (ZANU PF) when it was formed in 1963 with other nationalists such as President Robert Mugabe, Herbert Chitepo, Edgar Tekere and Enos Nkala. Born in 1920, Rev Sithole grew up in colonial Zimbabwe and like many of his contemporary black elites; he got a university education in the United States of America and soon became involved in Zimbabwean politics. Rev Sithole spent time as a teacher at mission schools such as Dadaya. The motto Mr Phiri Maseko learnt from Rev Sithole was to resonate throughout his life and proved to be very inspirational. His entry into politics later in his life was no doubt greatly influenced by Rev Sithole. His innovations for managing rainfall variability can also be said to have been inspired by the Rev Sithole who constantly reminded Mr Phiri Maseko to "work hard and fear nothing". Rev Sithole urged African people to work hard in everything they did and were not supposed to fear the colonial authorities but demonstrate that they could be high achievers.

Another important influence on Mr Phiri Maseko at Dadaya Mission was his agriculture teacher Mr Bunkam Ndlovu from Gwatemba, about 40 km to the west of Zvishavane. Mr Ndlovu studied agriculture at one of the country's pioneer agricultural colleges. The Agriculture teacher taught Mr Phiri Maseko about digging basins for planting fruit trees. Every Saturday morning Mr Phiri Maseko and his classmates watered the fruit trees. Mr Phiri Maseko realised that the basins stored water to secure the lives of the fruit trees.

After finishing Standard 6 in the late 1940s, he taught in rural Zvishavane for a few years. He then moved to Mt Darwin a small town about 140 km to the north east of Harare where he took up a post as a teacher. In 1950, his father passed on after a short illness. Mr Phiri Maseko was devastated with the news of the demise of his father and returned to Zvishavane. He recalls the last time he saw his father when the Todds had come to bid farewell the Phiri family to briefly visit New Zealand. Mr Phiri Maseko said his father foretold his own death:

When Sir Garfield bade farewell to my father, he (my father) replied that he would not see the Todds again. Sir Garfield said that they would meet again after returning from New Zealand. My father insisted that they would not meet again. My father then asked the Todds to take care of me.

It was during this time that he decided to build a house on the stand that had been allocated to him by the village head. It was also during this time that he started to farm as he assumed his new responsibility to look after his mother and siblings. Mr Phiri Maseko knew that to be productive and harvest enough grain to feed his mother and siblings, he had to find ways of adapting to climate variability in the dry agro-ecological region 5. He decided to engage in market gardening and sold vegetables such as tomatoes, kale, cabbages, onions and kale. He looked for employment in order to supplement his income from market gardening. In 1957 he became a firefighter at the then Rhodesia Railways (RR) now National Railways of Zimbabwe. At RR, he got involved in political activism. In the 1970s, he was under house arrest because of his involvement in the liberation struggle. In the 1980s after the country gained independence, he met Ken Wilson and Ian Scoones. More of this is covered in more detail later in this chapter.

Ken Wilson and Ian Scoones were a major influence on Mr Phiri Maseko's life in the 1980s. The two were British PhD students doing their fieldwork in rural Zvishavane and were mostly based at Mototi in the Mazvihwa area of Zvishavane. They lived at the home of Mr Cephas Mukamuri, a local educationist (this will be covered in more detail in Chapter 5). This was the beginning of a relationship that would blossom. Ken Wilson paid Mr Phiri Maseko regular visits and has documented his work over the years. Ken Wilson played a major role in marketing Mr Phiri Maseko to the outside world. Ken Wilson nominated Mr Phiri Maseko (according to Mr Phiri Maseko) for the prestigious National Geographic Society award for his leadership role in conservation. This exposure to the outside world caught the eyes of many other international organisations such as Ashoka. Ken Wilson also helped Mr Phiri Maseko to secure funds from the World Development Movement in Britain to pursue his dream of setting up an organisation that would help smallholder farmers in sustainable ways of water management and soil conservation. The World Development Movement at the behest of Ken Wilson invited Mr Phiri Maseko to England.

Ian Scoones (now a Professor at the Institute of Development Studies at the University of Sussex, UK) first met Mr Phiri Maseko in 1986. He made a regular 'pilgrimage' to Zvishavane to see Mr Phiri Maseko. Together with Ken Wilson, Scoones also assisted in helping fine tune Mr Phiri Maseko's methods of water harvesting such as the technical aspect of how smallholder farmers could strengthen their dam walls so that they would not collapse. Like Wilson, Scoones nominated Mr Phiri Maseko for the King Baudouin International Development Prize in 2004. Besides influence from Ken Wilson and Ian Scoones, Mr Phiri Maseko's skills in agriculture were honed by an agricultural institution.

Mr Phiri Maseko had an opportunity to attend Makoholi Research Institute in 1973. His innovations were made known to the Ministry of Agriculture who invited him to attend courses in land and water management. I suggest the invitation could have been meant to pacify Mr Phiri Maseko so that he refrained from engaging in political activities. Such were the tactics used by the colonial authorities to try to stem the tide of the revolution against colonial rule, which at the time was engulfing the country. This is in no way to belittle his achievements as a farmer that made him an ideal candidate to do the course. Makoholi is an agricultural research station situated about 270km to the south of Harare. It falls under the Department of Research and Specialist Services. Mr Phiri Maseko attended courses in poultry; apiculture, gardening and aquaculture at the agricultural institute (see also Witoshynsky, 2000). After his stint at Makoholi, Mr Phiri Maseko began to implement on his plot some of the things he had learnt. For example, he initiated apiculture and aquaculture projects. He also took courses in market gardening. Market gardening greatly consolidated Mr Phiri Maseko's status as a latter-day *hurudza* by improving his financial status, which in turn increased his adaptive capacity.

He has redefined *uhurudza* by encapsulating what Steinmetz (*TIME Magazine*, 2014: 11) terms "enviro-preneurship". An enviro-preneur engages in environmentally friendly agro-business practices. For example, Mr Phiri Maseko uses compost to add fertility to his soil to grow what he terms "healthy food". His water harvesting practices have helped to reduce gully erosion on his plot and its environs. This is evidenced by deepened contours on his plot that store water rather than drain it away.

Mr Phiri Maseko can also be said to be an enviro-preneur because water he harvests has created lush vegetation that includes reeds he occasionally sells to basket weavers. The vegetation has attracted wildlife such as birds that roost on the trees. His guinea fowls breed in the thick vegetation on his plot, which acts as a hideout for the guinea fowls from predators. He sold the eggs that were laid by the guinea fowls. However, Mr Phiri Maseko said that he owed his enviro-preneurship to God. In his own words, he prayed to God for "guidance, perseverance and love" to make his market gardening project prosper.

Mr Phiri Maseko said that spirituality was a major source of inspiration for him. It is in this background of his spirituality that he viewed water and soil marrying in holy matrimony. The holy matrimony is evidenced by a bumper crop, as Mr Phiri Maseko would say. To facilitate that holy matrimony between water and soil on his plot (after his dismissal from RR in 1966), he says he read the book of Genesis 2 verses 8 to 10:

> And the Lord God planted a garden in Eden, in the east; and there he put the man whom he had formed. Out of the ground, the Lord God made to grow every tree that is pleasant to the sight and good for food, the tree of life also in the midst of the garden, and the tree of the knowledge of good and evil. A river flows out of Eden to water the garden, and from there it divides and becomes four branches (New Revised Standard Version Bible, Old Testament, 1989: 2).

He also always quotes Matthew 7 verses 7 and 8:

> Ask, and it will be given you; search and you will find; knock and the door will be opened for you. For everyone who asks receives, and

everyone who searches finds, and for everyone who knocks the door will be opened (New Revised Standard Version Bible, New Testament, 1989: 5).

These verses were a revelation in Mr Phiri Maseko's life. In his own words, the words from the Bible 'oiled' his innovative mind. They inspired him to do great things. He said that he was a living testimony of the good God can do in people's lives. In later years, he spent most of his time at home ensconced in an armchair on the porch, reading his Bible (see Photograph 3).

Photograph 3: Mr Phiri Maseko reading his Bible at home

Christianity runs in Mr Phiri Maseko's blood. He was baptised as a member of the Church of Christ at Sivanga in 1937. His parents were both Christians, and so were a major influence in imparting Christian values to the young Zephaniah. He said, "Christianity brings dignity in a human being". He had great admiration for a minister in the Church of Christ named David Mkhwanazi. Mr Phiri Maseko said of Rev Mkhwanazi that he carried the weight of Christ wherever he went and was well respected. Mr Phiri Maseko would say that respect is earned; you must work for it. His spiritual beliefs were at the heart of his innovations for climate variability; however, this is dealt with more extensively elsewhere in this chapter.

Mr Phiri Maseko was (during my fieldwork) in the process of helping build a church on the rock outcrop overlooking his home (see Photograph 4 below).

Photograph 4: The church Mr Phiri Maseko is helping to construct

At church gatherings, Mr Phiri Maseko, besides preaching the word of God, also used such a platform to disseminate his farming ideas. Some members of the Church of Christ have heeded his advice and have started harvesting water. He says the teaching of Christ was about the brotherhood and sisterhood of humanity and yet in colonial Zimbabwe there was racial discrimination. The next section focuses on his 'trials and tribulations' at the hands of colonial authorities.

Advent of war: Arrests, torture and immobility

During the height of Zimbabwe's liberation struggle in the mid-1970s, Mr Phiri Maseko's fame as a *hurudza* attracted prying eyes of the protagonists (Rhodesian soldiers and liberation fighters) who both in a bid to establish hegemony in the locality - victimised Mr Phiri Maseko. He had cut his political teeth on his detention days at

Gonakudzingwa in southeastern Zimbabwe during an era when the decolonisation process in Africa had been set in motion. In the 1960s, most African countries gained independence but this was not to be the case in the then Southern Rhodesia. Mtisi et al., (2009), argue that in Southern Rhodesia the white settlers were determined to jealously guard their political and economic privileges but Africans encouraged by the 'winds of change' blowing across Africa, sought to attain independence. Mtisi et al., also state that:

> ...visions of the future led to a complex, and often violent power struggle as various forces sought to define and redefine the political, social and economic boundaries of the desired nation (Mtisi et al., 2009: 115).

It was against this background that Mr Phiri Maseko found himself involved in political activism because of racism. While at the RR, he was involved in a nasty incident with a white colleague. He recalls the incident:

> It so happened that I was deliberately subjected to an electric shock by a white colleague. Instead of apologising, he laughed at me uncontrollably and I beat him thoroughly. He began to cry loudly in order to alert other workmates. Some of my white workmates came running to investigate what had taken place. I explained to them the incident that had happened and they all laughed at him.

This incident led him to get involved in trade unionism while at the RR. Trade unionism was a latent way of expressing political activism. Some of the leading lights during Zimbabwe's liberation struggle emerged from the trade union movement. Mr Phiri Maseko was soon involved in political activities by organising demonstrations against the colonial authorities. Trade unionists articulated their marginalised status at the hands of the colonial authorities during their meetings. It was during one of these meetings in 1964 that Mr Phiri Maseko was arrested and detained for two years at Gonakudzingwa restriction camp in southeastern Zimbabwe with other detainees. It was at Gonakudzingwa that he met political heavyweights such as Joshua Nkomo. Nkomo was one of the founding fathers of independent Zimbabwe and earned the nickname 'Father Zimbabwe' for his liberation efforts. He was the president of Zimbabwe African People's Union (ZAPU) one of the first mass national parties in Zimbabwe. The years spent in detention could partly have helped hone his innovative skills. Nkomo (2001: 123) says that Gonakudzingwa was part of Gonarezhou Game Park and was very remote. There were as many as 3000 detainees at Gonakudzingwa over a period of ten years (Nkomo, 2001: 127). Detainees occupied themselves by engaging in various activities such as doing lessons in farming, history of the country and playing soccer. Nkomo says:

> We took control of our own lives, set up our own camp government and ran it as a practical course in democratic administration. The camp was run by the central committee, whose members acted as the chairmen of specialised committees for education, reception, hospitality and so on (Nkomo, 2001: 127).

In detention, Mr Phiri Maseko broadened his knowledge on farming. He met with many *hurudza* among the detainees and held discussions to do with farming practices from the different parts of the country. The restriction camp also hardened him. Restriction camps such as Gonakudzingwa were used by colonial authorities to try to stop the tide of mass nationalism. Detention was meant to break detainees mentally, emotionally and physically to destroy their revolutionary fervour. The conditions in detention camps were appalling. For instance, one of the heroes of the Zimbabwe's liberation struggle, Tekere (2007: 68) described Wha Wha (about 90 km to the north of Zvishavane), where he was detained, as a "snake park" because it was snake infested. Tekere indicates that they "slept on the hard floor...Once we found a cat's head in the sack of meat we were given to eat...our *nyemba* (cow peas) were infested with weevils" (Tekere, 2007: 58 – 68).

The detainees had to be very creative to communicate with each other (Nkomo, 2001; Tekere, 2007). Gonakudzingwa consisted of many camps, which were guarded by colonial authorities who prevented detainees from communicating to each other for fear of insurrection. To communicate, the detainees devised a plan. Nkomo points out that:

> It was the dogs (patrolling along the paths of the camps) that enabled us to keep in touch with other prisoners. Each camp put out food for the dogs at fixed, but different, times of the day so they made a regular circuit of the camps. We fixed up little pouches behind their collars, so messages passed from one camp to another (Nkomo, 2001: 133-134).

From the insights, above, it is evident that in the face of adversity, one must be creative to survive. This kind of creativity was an influence on Mr Phiri Maseko later in his life. He confessed that he learnt a lot from his experience at Gonakudzingwa. The hardships faced by detainees as described by Tekere (2007) solidified Mr Phiri Maseko's inner conviction. The detention was meant to discourage detainees from supporting the liberation struggle but had the opposite impact than intended because detainees such as Mr Phiri Maseko became more resolute in fighting for independence.

Mr Phiri Maseko was released from detention in 1966 and immediately expelled from the RR. In a ZWP report (1990) in response to a visitor, Ms Tricia Spanner, about what motivated him to start his water project, Mr Phiri Maseko writes:

> I told her what my aims are over this project in Zvishavane as a whole …how I started and why…I told her that it was after I had lost my job (at the Rhodesia Railways) due to politics and I was taken to Gonakudzingwa Restriction Area for two years. When I came from there I had no job…during this time I was married with six children…the children wanted food, clothes, and schooling…Since I had no job…how could I have food for my children? (ZWP report, 28 February 1990).

He married his first wife Lizzy in 1951 and the couple were "blessed" with ten children (seven boys and three girls). In 1989, he married his second wife Constance and they have three children, two girls and one boy who are all alive. His first wife passed on in 2001 and today from his first marriage, only five children survive. In a rural environment of this kind, one had to realise high yields of grain for one to be able to provide for one's family adequately. I suggest that this too was a factor that motivated Mr Phiri Maseko. Of his children, only two are smallholder farmers. The two are sons from his first marriage. One of them has been actively involved in water harvesting activities with his father. He was employed by ZWP as a field officer for about six years. He potentially is poised to continue with his father's legacy of harvesting water.

I suggest that Mr Phiri Maseko's Malawian origins partly shaped his survival strategies; he was also influenced by Shona worldviews and cosmologies having been raised and having lived in a predominantly Shona speaking area. It seemed Mr Phiri Maseko, a master at presiding over 'marriages', 'married' Shona and Malawian customs to adapt to the harsh environment that characterises rural Zvishavane. In Zimbabwe, people of Malawian origins mainly grow cassava. He set aside 0, 50 acres for cultivating cassava to help mitigate climate variability. His interaction with the local people helped shape his perception of soil and water. He referred to himself as *mwana wevhu* (a child of the soil). He used to farm fish on his plot and attributed this to his Malawian origins. He said people from Malawi eat a lot of fish perhaps due to their proximity to Lake Malawi. In other words, there are several sources of influence that can be attributed to Mr Phiri Maseko's innovations. Nevertheless, the tide of the liberation war in Zimbabwe temporarily halted his innovations.

African nationalists launched what was termed the 'Decisive Phase' of Zimbabwe's liberation struggle (1972-1979). After a lull in fighting in the late 1960s, the nationalists embarked on what they considered the final assault to attain independence. This period was characterised by a huge influx of freedom fighters into rural areas. They were operating from bases in Mozambique and Zambia. The freedom fighters sought food and shelter from the rural communities and cached weapons at places only known by a few trusted villagers. Mr Phiri Maseko, by virtue of his standing in the community as a *hurudza,* an accomplished innovative farmer and an opinion leader, was one of the villagers the freedom fighters enlisted as a confidant.

In 1976 freedom fighters, cached weapons at Mr Phiri Maseko's homestead, and this led to his arrest. In the early 1970s, he had been involved in a legal tussle with the colonial authorities over his cultivation of the wetland located below his homestead as it was forbidden to do so. He was arrested thrice and had to pay a fine of about ten pounds each time. During the third arrest, he invited the magistrate to his plot to see for himself the 'crime' he had 'committed'. The Land Development Officer (LDO) accompanied the magistrate. When they got to his plot, the magistrate was happy with the agricultural methods he saw, and discharged Mr Phiri Maseko. However, Mr Phiri Maseko had become a thorn in the flesh of the colonial authorities, and it did not come as a surprise to them when arms caches were 'discovered' at his home in 1976. He had become a victim of his own fame. The Phiri Maseko homestead had become a space for competing interests between the two protagonists, the freedom fighters on one hand, and the Rhodesian soldiers on the other. Mr Phiri Maseko was now in the protagonists' collective crosshairs. To the freedom fighters, winning Mr Phiri Maseko over to their side meant helping legitimise the war effort, for if, opinion leaders such as him supported the struggle then the majority would follow suit. On the other hand, the Rhodesian soldiers' calculation was that a heavy-handed approach in dealing with Mr Phiri Maseko would deter would-be "terrorist collaborators". The Rhodesian soldiers regarded the freedom fighters as "terrorists" and the villagers such as Mr Phiri Maseko as "terrorist collaborators" and were supposed to be dealt with ruthlessly.

The colonial authorities were notified of the cached weapons allegedly from an informer in Mr Phiri Maseko's village. Resultantly, the hounds were drooling at the door. Mr Phiri Maseko's homestead swarmed with soldiers of the Smith regime. They harassed him and his family. The 'discovery' of the arms caches led to his arrest in 1976 and he was taken to Zvishavane where he was tortured for a week by security agents of the Rhodesian government. He was then taken to Wha Wha Prison. When he was released from prison, Mr Phiri Maseko was put in leg irons at his home for about six months, and then placed under house arrest - a development that immobilised him for the next four years dealing a blow to his innovations. The Rhodesian soldiers hoisted the settler regime flag at Mr Phiri Maseko's home to try to instil fear in the villagers (see also Witoshynsky, 2000). The move was also meant to contaminate the relations between Mr Phiri Maseko and the freedom fighters. The torture immobilised Mr Phiri Maseko; his injuries were so severe that he walked with a limp. Here called the events of the 1970s:

> *Mwanangu* (My son) the Rhodesians wanted to kill me. The major architect of my suffering was Mr Shaw (the Zvishavane district administrator in the mid-1970s) ... he was a bad man. Mr Shaw came to my home several times and organised meetings with the villagers warning them not to associate with me because I was said to be a "terrorist collaborator". There are times when I almost broke down but the thought of my family made me to remain resolute. The torture made me determined to help liberate my country. In Shona, we say '*Sango rinopa waneta*' (The forest rewards those who persevere).

He was to remain under house arrest until the advent of Zimbabwe's independence in 1980. This era was a brief setback to Mr Phiri Maseko.

Picking up the pieces: A time of "extraordinary catharsis"

I suggest that the advent of Zimbabwe's independence in 1980 was a time of extraordinary catharsis for Zimbabweans. I borrow the phrase 'extraordinary catharsis' from Clapham (2012). Clapham suggests:

> The moment at which a liberation movement comes to power is normally one of extraordinary catharsis. Whether this takes the form

of sandaled fighters sliding into the capital city from the countryside, as the discredited remnants of the old regime flee or surrender, or a formal handover in the wake of a negotiated settlement and founding election, it has something definitive about it. The war-weary population is generally happy to accept that the long conflict is over, and – whatever misgivings they may have had about the victors when the war was undecided…The war itself may well have created major problems of dislocation…The new regime's own fighters need to be settled into the very different world of peacetime life; refugees have to find their way back, often to home areas shattered in the fighting, and start to re-establish a normal existence…There is little opportunity… to relax and enjoy the fruits of victory (Clapham 2012: 7 -8).

Trying to understand the implication of what a time of catharsis was for Mr Phiri Maseko, one must understand the meaning of Mr Phiri Maseko. To help me unpack the meaning of Mr Phiri Maseko, I use the analysis of Rankine (2015) in her article on what it means to be tennis goddess Serena Williams. Rankine says, "There is no more exuberant winner than Serena Williams. She leaps into the air, she laughs, she grins, she pumps her fist, she points her index finger to the sky, signalling she is number one. Her joy is palpable" (Rankine, 2015).

I am inclined to try to give the meaning of Mr Phiri Maseko in relationship to his plot (his 'Garden of Eden'). He smiled, he joked, he sang, he talked with his plants and he told stories about his formative years as a *hurudza*, how his experiments helped him to let go of his emotional ardour. Mr Phiri Maseko's way of releasing his emotions was through what one of his visitors called the 'Eden therapy'. This is about how working on his plot nicknamed the 'Garden of Eden' was a source of joy to Mr Phiri Maseko. After regaining his freedom in 1980, his way of releasing tension was to do the activity he enjoyed most, relentlessly experimenting on his plot in a bid to survive in a harsh environment. In other words, Mr Phiri Maseko exuded happiness when he was farming. Exuding happiness provided catharsis. For him, the time for catharsis was through his engagement with non-humans on his plot after his house arrest.

Maya Angelou captures Mr Phiri Maseko's time of catharsis, in her poem, "*Still I rise*", in the prologue to this book. She writes, "I rise, from a past that is rooted in pain" and this could reflect Mr Phiri Maseko's main objective, to 'rise' from a state of despondency caused by years of house arrest and be a food secure *hurudza*.

Catharsis was about letting go emotions about his experiences during the liberation struggle. The combined effects of brutal torture and house arrest had affected his innovations. His house arrest for four years meant he could no longer attend to his plot. Moreover, having been placed in leg irons for six months at his home curtailed Mr Phiri Maseko's movement. Survival strategies he honed over the years helped mould him into a resilient farmer. Faced with a vision of a dystopic future where you have effects of years of house arrest colliding with rainfall variability in semi-arid Zvishavane, it was time to pick up the pieces and continue from where he had left; there was no time to relax. Picking up the pieces was about the emotion, the 'Eden therapy' as he revived his pet project that had been lying comatose. Picking up the pieces meant taking from where he had left off before his house arrest. It meant reviving his plot. I suggest that for Mr Phiri Maseko, picking-up the pieces was a time to 'exorcise the colonial ghost'. It was time to release emotions about his ill-treatment during the colonial era and get back to work.

Mr Phiri Maseko returned to the cathartic farming he was so passionate about. There was no time to relax, the time was for work to survive in his variable environment. His survival skills are amply demonstrated by the following story he told me:

> During my days at Dadaya Mission, I disliked being away from home. I greatly missed my mother and friends. I also missed looking after my father's cattle, because what we enjoyed most was milking the cows and drinking the milk. I also used to harvest lots of wild fruit. When I started school at Dadaya all this stopped. I then thought of a plan. My friend David Mvuu and I would run away from school and go home most of the weekends. We were punished for running away from school. On Mondays, Sir Garfield Todd, the school principal, would come to our classroom and say, "Zephaniah Phiri and David

Mvuu come to my office". Sir Garfield Todd would then use a whip to beat us. In order to avoid being harassed and punished by prefects and being sent to Sir Garfield Todd for a lashing, I hatched a plan. I bribed prefects with presents of wild fruit and bread I brought from home (chuckles). Remember I told you that when my father came from Malawi, he worked at a bakery, so he used to bake for us at home. Back then in the 1940s, bread was a delicacy eaten by a few African families. The plan worked, and soon I became untouchable. No prefect would punish me from then on.

This is when Mr Phiri Maseko deepened his contours and adapted an innovation that helped to conserve water that he called 'infiltration pits'. It was also during the 1980s that he constructed small dams on his plot. However, this will be discussed in more detail in Chapter 3.

Storage of water in the dams meant that he could grow crops all year. He broadened *uhurudza* by incorporating, for example, more intensive market gardening. He mainly grew beans, tomatoes and kale. During his heyday in the 1980s and 1990s, Mr Phiri Maseko supplied vegetables to places as far as 30 kilometres from his home. Some of the vegetables were cooked, dried and preserved as *mufushwa* by his wives. Preserving food such as *mufushwa* helps farmers build resilience to erratic rainfall. *Mufushwa* can be preserved for three years and Mr Phiri Maseko, besides consuming it with his family gave it to needy members of society, especially in the event of failed harvests. He asserted that *uhurudza* implies disseminating ideas that help other farmers to be productive. However, as he always said, "charity begins at home". Mr Phiri Maseko asserted that the hallmark of an enterprising farmer is to first demonstrate his/her *uhurudza* at home.

After his house arrest, his dream for creating a 'Mini-Garden of Eden' was realised. Mr Phiri Maseko said that the water he harvested created lush vegetation on his plot, a 'Mini-Garden of Eden' home to a variety of birds and small animals. The garden is convivial with other elements of nature and several animal and plant species can be found here. Mr Phiri Maseko encapsulated the idea that if one is convivial with nature, one also "rhymes" with other elements of nature as suggested by Professor Francis Nyamnjoh (2013) during a discussion we held in his office. This is evidenced by the example of a stray zebra that has made Mr Phiri Maseko's homestead its new home. Mr Phiri Maseko had this to say about the stray zebra:

My son, this zebra (see Photograph 5) rhymes with me, it decided to make my home its own home as well. Imagine this, there are many people with donkeys in Mapirimira Ward, but it (the zebra) chose the home of a *mubwidi* as its home. You notice there is a goat that runs away from its owner and comes here every day; Aah Zephaniah, I do not know what to say. If you also look at the attic you see those pigeons perched, they abandoned their owner and have since made this home theirs. I also told you about the python we saw here at this very homestead. So, you see my son, I do not just rhyme with other humans; I rhyme with nature as well. Zephaniah *bakithi* (laughs).

Photograph 5: The zebra at Mr Phiri Maseko's homestead that lives with his donkeys

From the 1980s and thereafter, Mr Phiri Maseko encapsulated the latter day hurudza. This is evidenced by how he did extension work by giving advice to visiting farmers. He chided the government demonstration workers when addressing his visitors (see Photographs 6, 7 and 8). Addressing a group of farmers that had visited him from neighbouring Chivi District in October 2012, he said:

> Why are your extension officers not here? Why did they send you here? Are they not the ones who are supposed to be doing extension work? Please next time you come (here) I want you to bring the extension workers so that they come and hear my message as well.

Photograph 6: Mr Phiri Maseko (seated next to a sand trap) gives a lecture to visiting students from Rio Tinto Agricultural College

Photograph 7: Mr and Mrs Phiri Maseko pose for a photograph with visiting teachers and students from a high school in neighbouring Shurugwi District

Photograph 8: Mr Phiri Maseko addresses visitors from the United States of America and the United Kingdom

The visitors at Mr Phiri Maseko's plot appreciate the importance of his ideas especially given that they say that visits by extension officers are rare. He used that opportunity of meeting his visitors to market his agricultural practices. The agricultural practices have made him famous as awards were 'pouring in' non-stop.

Awards: A canary in a coal mine

The awards Mr Phiri Maseko won are a stark testimony to his vision that with increased rainfall variability, harvesting water is an effective pathway to help secure rural livelihoods. His vision in many ways acts as an early warning system (just like what a canary did in a coalmine) to the dire effects of a changing climatic environment. Thus, he was like a canary in a coalmine. This vision did not escape the recognition of many international organisations.

"Mubwidi akazviita" (A person of Malawian origins won awards), boldly declared Mr Phiri Maseko holding his National Geographic Society certificate. The term *mabwidi* (plural for *mubwidi*) resonates with the term *makwerekwere,* used in South Africa to refer to foreigners (Mabeza, 2013: 130). Mr Phiri Maseko became lively each time he talked of his awards and how he reaped international recognition. Despite having been born in Zimbabwe and living among the Shona, he jokingly referred to himself as a *mubwidi*, a poignant reminder that even today there are still prejudices against people of Malawian descent in Zimbabwe. He was in fact referring to the predicament of the 'mobile Africans' (this includes his father) and how they were (and even today) ill-treated by some of the locals. Mr Phiri Maseko might not be considered a 'mobile African' from Malawi but among the locals, the name Phiri Maseko conjures an element of being foreign. To many locals, the name Phiri Maseko is associated with Malawian descent and hence some locals may consider him 'mobile' and refer to him as a *mubwidi*. Nyamnjoh (2013a: 653) says that "mobility is more appropriately studied as an emotional, relational and social phenomenon as reflected in the complexities, contradictions and messiness of the everyday realities of encounters informed by physical and social mobility". Mr Phiri

Maseko was 'domesticated' by the locals in the same way he 'domesticated' water. To gain recognition among your hosts as a mobile African, you must excel in whatever you do (see also Nyamnjoh 2013a).

Mr Phiri Maseko carved out a niche in the country's hall of fame through his adroit water harvesting skills. He was a globally acclaimed water harvester as evidenced by the international awards accorded him. These include the Ashoka Fellowship in 1997 and the National Geographic Society Award for Leadership in African Conservation in 2006, when he picked up a king's ransom in prize money. Mr Phiri Maseko was awarded the Ashoka Fellowship for his concept in innovative water harvesting for farmers in southern Zimbabwe. Ashoka conferred upon him the award also as recognition of how his ideas have not only been implemented in Zimbabwe but have spread to countries such as Malawi, Uganda and Zambia.

Ian Scoones nominated Mr Phiri Maseko for the 2004/2005 King Baudouin International Development Prize as already alluded to. Although Mr Phiri Maseko did not win the prize, Scoones posits that:

> I harbour no doubt that the contributions of this candidate fully meet, indeed exceed, the desired qualities stated by the King Baudouin Foundation with regard to an individual whose life work has been dedicated to "sustainable achievements in improving the lives of people in the developing world". Moreover, the "multiplier effect" of Mr Phiri's life work resonates fully with the Foundation's goal to reward the "opportunities they (nominees) give to the people they serve to take control of their own development" (extract from Ian Scoones' nomination letter to the 2004/2005 King Baudouin International Development Prize, 28 January 2004).

The 2006 full award citation by the National Geographic Society (see Photograph 9 of Mr and Mrs Phiri Maseko with the National Geographic Society certificate and atlas) applauds him:

For exemplary leadership in community-based resource conservation among dryland farmers in Zimbabwe and across southern Africa;

For fifty years of pioneering research, experimentation and refinement of innovations in rain water harvesting, soil conservation and dryland farming;

For founding the Zvishavane Water Project;

For ceaseless dedication to teaching the importance and benefits of conservation;

For inspiring farmers, practitioners, scholars, and the development of new conservation programmes;

For harvesting rain for school water supplies and for restoring watersheds;

For humanitarianism in giving water, seeds, and knowledge to nurture them;

For wisdom and courage in working to improve the wellbeing of fellow farm families through resource conservation for food security; and

For lifelong altruism, that enriches our human spirit (National Geographic Society, 2006).

In addition, the Britain Zimbabwe Society (BZS) in 2007 said that Nigel Hulett of Granadilla Films produced a documentary on Mr Phiri Maseko, entitled *The Water Harvester*, based on the book with the same name written by Mary Witoshynsky.

Mr Phiri Maseko was invited to many fora in countries such as Kenya and Ethiopia to give lectures on water harvesting. In 2012, Mr Phiri Maseko and his wife Constance attended a regional Environment and Sustainability Leadership Workshop from 7 to 9 November, in Howick, South Africa. The invitation to the workshop preceded a visit to his plot by officials of the Southern African Development Community Regional Environmental Education Programme (SADC REEP), based in Howick. He gave a good account of himself by chronicling his humble beginnings in water harvesting as a smallholder farmer in semi-arid Zvishavane.

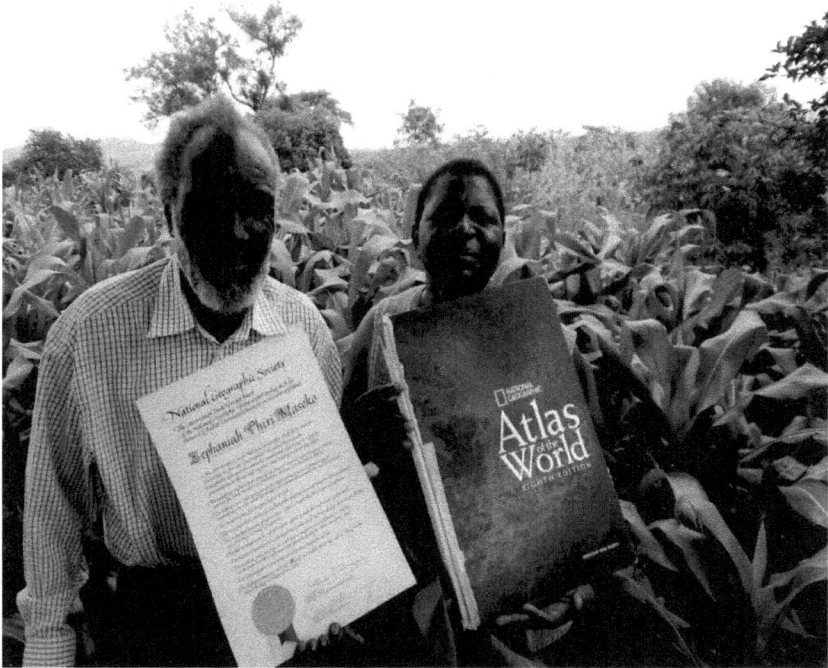

Photograph 9: Mr and Mrs Phiri Maseko pose with the National Geographic Society certificate and atlas

In 2012 the idea for the Phiri Award for Food Sovereignty was conceived to "...honour the lifetime work of Zephaniah Phiri Maseko and his family by recognising the efforts of exceptional small-scale farmers who innovatively spread the cause of food sovereignty and family farming in Zimbabwe" (Wilson, 2012).

John Wilson, the brain behind this award, was instrumental in the formation of an organisation in Zimbabwe called the Participatory Ecological Land Use Management (PELUM). PELUM, which facilitates sharing and learning events has spread its operations across eastern and southern Africa and has taken on board the Phiri Maseko water harvesting techniques as part of its message to other smallholder farmers.

On 24 October 2014, the Phiri Award for Farm and Food Innovators was launched in Harare. The award is meant to help identify local innovators so that they share their innovations, technologies and sustainable farming practices with other farmers

(Scoones, 2014). It is hoped that the award will promote local innovation in Zimbabwe (Scoones, 2014). Interestingly, one of the inaugural winners of the award is one of Mr Phiri Maseko's adopters, Mr Mawara from Mazvihwa in Zvishavane. Mr Phiri Maseko spent time in Mazvihwa doing extension work during his days at ZWP. I write about this more comprehensively in Chapter 4. He continued to receive recognition for his innovations on managing climate variability and deservedly so. He bought assets that include a car (see Photograph 10), using proceeds from his awards, which he converted into a taxi to try to diversify his livelihood strategies. This intervention can be viewed as transformational in relationship to Rickards and Howden (2012)'s perspective on seeking out occupation which supplements Mr Phiri Maseko's agricultural activities.

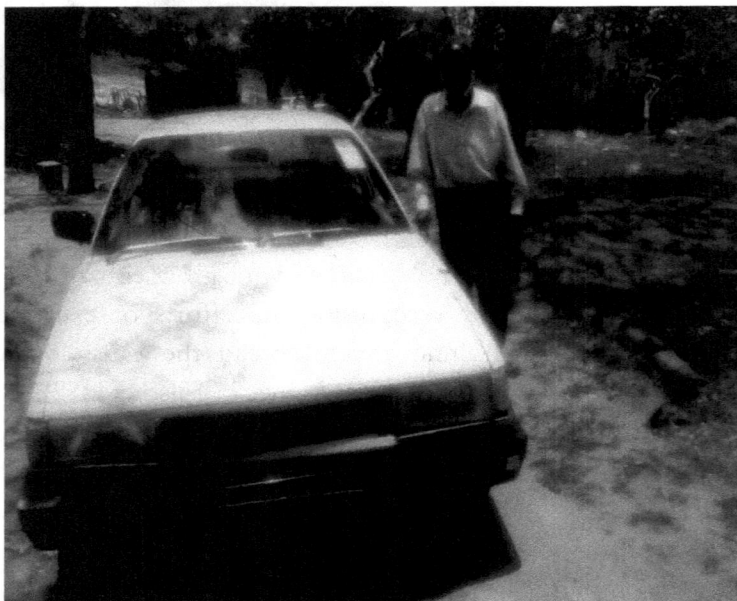

Photograph 10: Mr Phiri Maseko standing next to his car, bought by proceeds from his uhurudza

Colleta Chitsike, a visitor at Mr Phiri Maseko's plot in July 2010, penned an ode to *Va*Phiri (Mr Phiri Maseko) called *"The Water Conservation Visionary"*:

To all living things
Water is the same
*Va*Phiri's wisdom
In all languages
Water is the same
*Va*Phiri's wisdom
Water scarcity is the future problem
*Va*Phiri's wisdom
Water rate of evaporation
*Va*Phiri's wisdom
Water and management of land resources
*Va*Phiri's wisdom
Rate of soil erosion
*Va*Phiri's wisdom
Soil moisture and availability
*Va*Phiri's wisdom
Ah, Zephaniah, *bakithi!*

She writes a postscript (Wilson, 2010): "I remember *Va*Phiri's wisdom. The man who talked and warned about water and climate change long before it was on the development agenda". Such was the appeal of Mr Phiri Maseko whose popularity saw him playing an influential role in the local chapter of the ex-detainees' association.

Mr Phiri Maseko was a member of the Zimbabwe Ex-Political Prisoners Detainees and Restrictees Association (ZEPPDRA). This organisation caters for the welfare of ex-detainees of Zimbabwe's liberation struggle. The other members of ZEPPDRA in Mapirimira Ward are I Sibanda, T Dinga, I. Magasa and S Mketiwa. They met at Mr Phiri Maseko's home for monthly briefings. The ex-detainees felt they should meet at Mr Phiri Maseko's home because he could no longer walk far because of the injuries he sustained at the hands of the Rhodesian security agents. The other reason could be that they valued him as a charismatic community leader whose influence could be of benefit to their organisation. Ex-detainees in Zimbabwe are entitled to a monthly allowance but it appears to be a pittance given the country's volatile economic environment. Allowances as these may be inadequate for smallholders to build adaptive capacity in a changing climatic environment.

Building adaptive capacity

Intrinsic to Mr Phiri Maseko's ability to adapt to climate variability was his strong adaptive capacity. Two characteristics are particularly prominent: conviviality and creativity. His creativity can be seen in the way he fused local and external knowledge to complement each other. For example, his method of preserving groundnut, cowpea and pumpkin seeds by use of herbs he learned from his father. His preservation of seeds helped him to refrain from monoculture, and he purchased maize seed from seed houses.

Examples of his conviviality included his version of the local practice of *kutema ugariri*. Given that some of his adopters worked for him in return for technical advice implies he realised that local practices are important (and that they are working or have traction with these other smallholders) in his quest to build adaptive capacity. The adopters involved in *kutema ugariri* helped him in market gardening where they mainly produced such crops as cabbages, kale and tomatoes. Mr Phiri Maseko used local methods of drying the vegetables to preserve them for times of scarcity.

This chapter traced the genesis of Mr Phiri Maseko's creativity. I argue that creativity was important in reducing his vulnerability. Such innovations as sand traps and deepened contours - a result of his creativity - harvest water that helped him to be less vulnerable to effects of climate variability. Mr Phiri Maseko realised that survival in dryland Zvishavane hinged on creativity and being able to study local dynamics to construct innovations that would help him to harvest water he needed for his crops.

His trailblazing research on water harvesting helped in moulding him into a *hurudza*. *Uhurudza* is about building adaptive capacity. He redefined the *hurudza* concept of Shona societies. For Mr Phiri Maseko, *uhurudza* meant enviro-preneurship. Enviro-preneurship has potential to play a vital role in building wealth thereby reducing smallholders' vulnerability to a multi-stressor environment. Wealth that Mr Phiri Maseko accrued due to his *uhurudza* includes livestock. His livestock comprised cattle, goats and donkeys. He used donkeys as draught power. The ability to rear appropriate livestock in a semi-arid region was another way he managed to develop strong adaptive capacity.

Mr Phiri Maseko reduced his own vulnerability because of his adaptive capacity that he built through his access to social and financial networks. He won many awards with monetary value, and created wealth for him. Mr Phiri Maseko's wealth also emanated from salaries he earned from ZWP and per diems at international conferences, forums that helped him gain recognition for his water harvesting techniques. This recognition fostered social networks that linked him to international scholars and practitioners who brought further resources, both knowledge/intellectual and financial leading to his innovative agricultural practices. Rodima-Taylor et al., agree that:

> ...informal, formal, endogenous and externally initiated institutions are interdependent and equally important in the process of adaptation and innovation (Rodima-Taylor et al., 2012: 111).

Mr Phiri-Maseko was linked to international networks such as researchers Ken Wilson and Ian Scoones who assisted him with ideas on some of his projects. Both Wilson and Scoones nominated Mr Phiri Maseko to prestigious awards. The awards are spin-offs of Mr Phiri Maseko's development/professional/innovation trajectory. Mr Phiri Maseko was on a trajectory that spanned five decades of 'marriages'. The 'marriages' reflect the concept of conviviality that helped him boost his adaptive capacity.

Conclusion

The very "stone that the colonial authorities harassed, abused and rejected has metamorphosed into the chief cornerstone" of smallholder farmers' innovations for managing climate variability in semi-arid southern Zimbabwe, according to Mr Phiri Maseko. He rose from a "past rooted in pain" as Maya Angelou would say. Mr Phiri Maseko assumed the role of a community leader in Ziyabangwa Village and Zvishavane District. He always hosted many people who came from far and near for advice, mostly on issues to do with his water harvesting techniques. From Mr Phiri Maseko's Visitor's Book, Kevin Lowther (CARE, US) concurs: "This is the beacon of light for the region. I learned much" (see also Wilson, 2010). Mr Phiri Maseko encapsulated *uhurudza* of Shona communities. He redefined *uhurudza* to include climate variability awareness and extension services. Awards for his inspiring work on water harvesting have provided a new dimension to the *uhurudza* concept. He demonstrated himself to be indomitable and hugely charismatic as a *hurudza*. Thus, 'marrying' suaveness and steel endeared him to the people of Zvishavane and beyond.

This chapter has shown that Mr Phiri Maseko's innovations are about conviviality as a practice of relationship building, and that this is deeply grounded in a widespread African philosophy. Therefore, as an intervention in contemporary scholarship on agriculture, development, and the anthropology of climate change, Mr Phiri Maseko's work enables the beginning of a decolonising dialogue, one that is grounded in regional philosophy and practice. These principles are widely reflected in regional thought/philosophy and conviviality as a practice of building relationships shown in this chapter not only between people (*hurudza*, everyday generosity) but also between people and soil and water. This is not only limited to Shona thought but also wider literature in terms of thinking relationally such as Wangari Maathai (2006) and the "foresters without diplomas" through their reforestation efforts in Kenya. In addition, the story of Robert Mazibuko, a South African farmer who saw the value of soil in its relationship to people's livelihoods (Bloch, 1996) adds to the regional philosophy about land.

The next chapter will explore his innovations for managing climate variability and their importance. The chapter puts into perspective his innovations that illustrated his arduous journey to an award-winning *hurudza*. The chapter further bolsters the study's argument that smallholder farmers' innovations that transcend barriers are vital for building resilience in semi-arid regions, as demonstrated by Mr Phiri Maseko. Therefore, 'Thou shalt not only survive adversity, but thrive in it' if we embrace smallholder innovations, what this study calls 'new tricks'.

Chapter 3:

'New tricks':
Managing rainfall variability

Zvinhu zviyedzwa chembere yekwaChivi yakabika mabwe ikaseva muto (It is important to experiment for an elderly woman in Chivi, rural Zimbabwe cooked stones and ate the soup) Shona proverb.

Individual and societal adaptation to climate is nothing new, neither as an empirical reality nor as a theoretical construct. The resource irregularities offered by different climates and the precariousness which emerges from the vicissitudes of climate have both acted as significant stimuli throughout human history for social and technological innovation (Adger et al., 2009: 336).

Introduction

The Shona proverb, *zvinhu zviyedzwa chembere yekwaChivi yakabika mabwe ikaseva muto* (It is important to experiment, for an elderly woman in Chivi [in rural Zimbabwe] cooked stones and ate the soup) reinforces my argument that innovation is characterised by trial and error. The elderly woman from Chivi was starving and experimented with cooking stones. She was surprised by her level of creativity when she 'ate' 'soup' from stones. Mr Phiri Maseko exhibits a "rare mix of relentless tenacity and technique" (to borrow a phrase from Anwar 2014: 1) as demonstrated by his innovations. From the many stories, he told me during my fieldwork, his innovations reflect an innovator with immense reserves of strength and determination. The tenacity and desire with which Mr Phiri Maseko strove to find seemingly elusive solutions to managing climate variability made him a hero, thus earning him recognition for innovative agricultural practices of water harvesting in semi-arid southern Zimbabwe. He built structures that help build the resilience of his surrounding ecosystem by harvesting water. Harvesting water has helped to raise the water table on his plot ensuring that he has enough water for watering his crops and domestic use. Because of the availability of water in his "water plantation" - his plot - he implemented a wide array of agricultural practices to build resilience including – market gardening, crop diversification, crop rotation, boosting soil fertility by adding manure, apiculture and aquaculture. It is these innovations, a breath of fresh air that this study labels 'new tricks'. These are the 'new tricks' that this study (in the opening line of the Introduction) says the establishment in the development discourse seem not ready to embrace. The 'new tricks' are a bold attempt and departure from the "usual-usual of development." They are the 'Phiri Maseko agricultural ecology'.

Drawing on the Shona conceptualisation of water, soil and marriage, this chapter offers an analysis of the 'Phiri Maseko agricultural ecology' characterised as it is by the use of relational metaphors

Disintermediation: The Phiri Maseko ecology

Extraordinary circumstances warrant extraordinary solutions. To that end, Mr Phiri Maseko, arguably one of the most disintermediated smallholder innovators in rural Zimbabwe, 'dumped' extension workers, the traditional middlemen, thus, rejecting conventional wisdoms in his quest to adapt to an increasingly variable environment. In an article in *Time Magazine* (January 2016), Von Drehle defines disintermediation as "dumping the middleman". The term disintermediation gained prominence with the advent of the Internet Age which has witnessed the downsizing of major retailers as customers buy online directly from manufacturers (Von Drehle, 2016). Mr Phiri Maseko, clearly a rebel with a cause, was on his own trajectory. He sidestepped extension workers, the government's 'boots on the ground' and initiated his own unique innovations. In rural Zimbabwe, the mandate of extension workers is to give technical advice to smallholder farmers.

The Phiri Maseko ecology therefore is about Mr Phiri Maseko rebelling against the status quo. His agricultural practices (new tricks)

are testimony to his disintermediation efforts. Mr Phiri Maseko's marriage of water and soil, as a water management system is central to his farming practices. Water and soil, the basis of human existence, are so valuable, they are reflective of a holy matrimony hence they are associated with sacredness, as evidenced by Shona terms such as *pasichigare* and *dzivaguru* (Moyana, 1984: 13). *Pasichigare* refers to the valorisation of sacred sites; for example, *dzivaguru* is a water pond that is sacred because it might be the dwelling of mermaids. The *pasichigare* concept is further illustrated by the significance of burying the *rukuvhute* (umbilical cord) in the soil. When I walked his land with him, Mr Phiri Maseko would usually smile and remind me that his *rukuvhute* lies in that soil. This connection ensures a lifelong attachment to the soil in a particular area and for many, that soil feels sacred. To rupture that relationship may result in 'broken marriages' which in turn have adverse effects on crop production on one's land. The underlying principle of 'broken marriages' is therefore that if soil and water go separate ways, it would lead to soil erosion and gully formation, as well as impacting on whether smallholders were able to produce enough for their families: that in turn would impact on well-being, fertility, and child health. Mr Phiri Maseko would say that the idea that 'broken marriages' lead to a cease in production infers a relationship between the moral realm and the productive capacity of the soil. This means that central to his philosophy is the stewardship of the land. He says, "If you take care of the land, the land will take care of you". His philosophy of water and soil resonates with Klein's critique of what she calls the 'astronaut's eye worldview' – which she suggests is:

> ... just looking down at Earth from above. I think it's sort of time to let go of the icon of the globe, because it places us above it and I think it has allowed us to see nature in this really abstracted way and sort of move pieces, like pieces on a chessboard, and really lose touch with the Earth. You know, it is like the planet instead of the Earth (Klein, 2013).

Mr Phiri Maseko's recognition of the interdependencies between humans and nature, and giving respect to non-humans, is how he turned the 'astronaut's eye worldview' on its head. It is recognition that humans should not place themselves above nature but rather that there ought to be a symbiotic relationship for humans to realise benefits. Turning the 'astronaut's eye worldview' on its head was a difficult journey for Mr Phiri Maseko. This worldview compelled him to construct water-harvesting structures starting from the1960s. Ultimately, Mr Phiri Maseko's technical nous paid dividends, with his innovations to climate variability transforming his plot.

The Phiri Maseko ecology also resonates with the Shona proverb: *Sango rinopa waneta*, (which means the forest rewards those who persevere). The Phiri Maseko ecology is a strategy, which enabled him and others to survive in a region with an average of about 546mm of rainfall per annum. In a published interview, Mr Phiri Maseko says: "Sure, it's a slow process, but that's life. Slowly implement these projects, and as you begin to rhyme with nature, soon other lives will start to rhyme with yours" (Lancaster, 1999). For Mr Phiri Maseko, the concept of rhyming with nature in this section is a crucial one, as it refers to the importance of appreciating that the survival of people and households depends on a healthy socio-ecological system. In other words, if humans degrade the environment then the ecosystem will be unable to offer the much-needed ecosystem services that ensure human survival. Water is one such vital element that life on earth, if deprived of it, will cease to function.

In terms of these philosophies, Mr Phiri Maseko harvested water and distributed it on his plot with a variety of infrastructures and inventiveness, most notably - the deepened contours - contrary to advice from extension workers. The following section discusses his innovations.

"Slow it, spread it, sink it, store it, and share it": Structures for harvesting water

> The complexity, immediacy, and ubiquity of environmental problems and crises demand novel and unusual human responses (Agrawal and Lemos, 2007: 39)

Innovations that 'slow', 'spread', 'sink' and 'store' water that is then 'shared' by his neighbours have turned environmental challenges into opportunities for smallholders such as Mr Phiri Maseko. Such types of innovations are a way of managing and spreading risks of climate variability. Adger says, "It is argued by ecologists that resilience in natural systems provides the capacity to cope with surprise and large scale changes – this is precisely what will allow innovation, coping with change… (Adger, 2000: 361).

Mr Phiri Maseko described his innovative agricultural system characterised by water harvesting based on a deep understanding of the types of soil in semi-arid Zvishavane - *musheche* (sandy soils) and *chinamwe* (clay soils) and a bit of hydrology of how *makuvi* (wetlands) according to Wilson:

> What happens in natural wetlands in this environment is that water flowing from granite hills and sandy soils aquifers comes up against accumulations of clay or impermeable rock areas down-slope and rises to the surface as springs and wetlands. These *makuvi* provide rich areas for grazing, excellent well sites and productive farming areas for wetland crops during the rains and vegetables in the dry season…What *Va*Phiri (Mr Phiri Maseko) discovered he could do was more than these traditional uses: he found that he keeps more of the water in the wetland during the rainy season than occurs naturally (Wilson, undated: 3).

His plot is composed of mostly sandy, loamy and clay soils. The upper part near the homestead is mainly made up of sandy soil and the lower areas are clay soils. The sandy soils allow infiltration of water and clay soils hold much of the water from escaping the plot.

This combination of soil and water enables a management system drains the vlei enough to cultivate it when wet and recharges it enough when dry. Mr Phiri Maseko made the heavy clay soils on his plot manageable by incorporating organic matter to increase its fertility hence transforming it and creating productive soils. The donkey pump (see Photograph 1) illustrates how much the soil level has risen. From the base of the pump when it was installed in the 1970s: by my measurement, there has been a rise of about 15cm. In keeping with his approach of harnessing the propensities of every situation, this pump for many years was powered by a donkey, which Mr Phiri Maseko motivated (with success) by placing maize cobs along its orbit.

Photograph 1: Spreading water - Mr Phiri Maseko demonstrates how the now disused donkey pump operated

He said he observed the water as it flowed down a *ruware* (rock outcrop) near his home and realised that when rain fell and was not slowed, there was very little infiltration. He then decided to experiment and constructed a sand trap at the foot of the rock outcrop next to his home and below that, he planted crops. He also planted crops where there was no sand trap. After a short while he observed that crops below the sand trap where growing much better as compared to crops planted where there was no sand trap. Mr Phiri Maseko also noticed that soil in hollows where water gathered after a storm remained wet. With these observations in mind, he decided to construct longer sand traps for the purposes of slowing down the flow of the water. As the water cascades down the rock outcrop, it is slowed by the sand traps he constructed to facilitate infiltration. These sand traps (see Photograph 2) also prevent soil erosion from taking place since they minimise the erosive power of the water by reducing its velocity.

Photograph 2: Slowing water - Mr Phiri Maseko's sand traps that slow down flow of water

When the water reaches sand traps most of it sinks into the soil. As the water infiltrates into the soil it is 'stored'. To construct the sand traps, Mr Phiri Maseko used the resources at his disposal including stones, shovels, hoes, mattocks, pick axes and an ox-drawn cart. He also harnessed the manual labour of his family and some of his neighbours to help construct the sand traps, check dams and canal. Research has shown that a construction-intensive "engineering approach" to water harvesting, which uses "large quantities of expensive bought-in materials is not a viable answer for agricultural development" (Everson et al., 2011: 149). Mr Phiri Maseko's water harvesting was premised on use of cheap materials as indicated above. "Slow it, spread it, sink it, store it, and share it" is the dictum that summarises the thinking behind his interventions.

Mr Phiri Maseko's observations influenced his constructing the 'immigration centre' for water to move in to the soil at the foot of the rock outcrop. The 'immigration centre' was constructed in 1969 (see Photograph 3) and its purpose is to keep water. He termed it the 'immigration centre' because without it water would 'migrate' (run-off) with soil causing soil erosion but with the advent of the 'immigration centre', the water 'checks in' at the 'immigration centre' where he welcomed it and 'told' it to sink gradually into the soil (see also Lancaster, 1999). While in part, this might be a sly wink at immigration officials at border posts to impress on them to be hospitable, it also reflects his wider practices of conviviality, since in the same way he welcomed strangers and guests, he welcomed water on his plot and directed it to go where it would be comfortable.

When he advised other smallholder farmers, he emphasised the need to 'welcome' water on their plots to help reduce the incidence of soil erosion, or alternatively, he drew on the metaphor of marriage, advising other smallholder farmers to harvest water in order to refrain from 'broken marriages' on their plots where water is not harvested and instead washes away soil causing erosion. A 'broken marriage' leads to crop failure due of lack of water and fertile soil, and is thus unproductive. In other words, 'broken marriages' were due to bad water management.

After the water fills the 'immigration centre', it moves down to Mr Phiri Maseko's fields as seepage. Water that flows into the 'immigration centre' is partly drained by pipes (see Photograph 4) and the rest is run-off water that cascades down the rock outcrop. The 'immigration centre' ensures the adequacy of water that he harvested for his crops. He said that if it filled thrice during the rainy season, he would have harvested enough water for his crops to see him through to the next agricultural season. I did not have the opportunity to witness the 'immigration centre' fill with water due to the low rainfall amounts that fell in the 2011/2012 and 2012/2013 agricultural seasons. However, I heard that it filled up once when I was attending engagements elsewhere in Zvishavane during my fieldwork.

Photograph 3: Sinking water - the 'immigration centre' that sinks water into the soil

Photograph 4: Spreading water - pipes that harvest water from the rock outcrop and drain it into the 'immigration centre'

Besides harvesting run-off water that cascades down the rock outcrop, Mr Phiri Maseko's other innovations harvest water at his homestead: these innovations are also integrated into a language of relationships. His first innovation was what he called the 'selfish tank' (see Photograph 5) which collects water harvested from the roof of his house. However, he was not happy with this tank because it was 'selfish' since it was only his family that benefited from it. He said his fruit trees did not benefit from it as well. Thus, he also constructed an underground tank that he called the 'social tank' or alternatively the 'poor man's tank'. He called it the 'poor man's tank' since anyone can construct this type of tank by using readily available resources such as disused metal sheets and stones rather than the costly alternative of bricks and cement. He first dug a pit in the ground, about one and half metres in diameter and two metres deep. He did not use mortar but used stones piled one on top of the other from the bottom of the pit going up. This enables trees to benefit from the water in the tank. On top of the tank, he put a metal sheet that he covered with soil. The underground tank conveys the message about the need to save water, a scarce resource in the areas. This is the message he imparted to his visitors.

There is no water that is wasted at Mr Phiri Maseko's home. Even dishwater (see Photograph 6) is drained into the 'social tank'.

Some of the water harvested by structures located near his home andwater that is harvested by his structures at the foot of the rock-outcrop moves through the soil to his out-fields a few metres away.

The canals are deepened contours (see Photograph 8): the terms are used here interchengably. The deepened contours rank among Mr Phiri Maseko's most important innovations because of their capacity to simulteneously store water and move it across the plot. In deepening contours, he was motivated by his observations of colonial contours and how they fostered gully erosion.

Photograph 5: Mr Phiri Maseko's daughter-in-law fetches water from the 'selfish tank'

Photograph 6: Sinking water - dishwater is drained into an underground tank, the 'social tank' through the white and black pipe at the base of the trough

The canals are partitioned by clay walls through which are inserted pipes for distributing and managing water. These pipes drain excess water from the fields into the canals. The additions to the canals prove that Mr Phiri Maseko was not a one-idea man as Wilson (undated) would say, he continuously experimented. In the 1990s, he inserted small pipes into the walls that partition the canals (see Photograph 7). This he did to help drain water from one partitioned part of the canal back to where it had come from. This is a way of managing feedback and helps to ensure that water is adequately distributed on his plot. He planted trees and vetiver grass to hold the soil from collapsing into the canals and reduce the rate of evaporation.

Photograph 7: Spreading water – pipes inserted into clay walls that partition canals on Mr Phiri Maseko's plot

Along the deepened contours (see Photograph 8), Mr Phiri Maseko dug small ponds that store water. He called them 'infiltration pits'. One of his visitors re-named them the 'Phiri Pits'. They harvest run-off water. Ken Wilson writes:

Va Phiri (Mr Phiri Maseko) did not like to seewater gathering up and gaining force, and then escaping fields, causing erosion and drying in the process. Therefore he adapted a pitting system he had heard about from other innovative farmers and began in 1987 to constuct pits to capture and enable infiltration of heavy rainfall...and gradually extended the practice down the fields through the 1990s...he has also built 'Phiri Pits' to capture water draining from the road...(Wilson, undated: 3).

Photograph 8: A deepened contour

To help absorb major flooding on his plot, Mr Phiri Maseko constructed dam spillways so that excess water is drained from the plot. The spillway helps in reducing waterlogging in years with high rainfall amounts. He also planted groves of banana and sugarcane (see Photograph 9). These plants use a lot of water, but ensure the plot is well drained.

He also constructed three small dams from the early 1970s to the1990s. The clay dam on the lower part of the plot holds the bulk of the water on the plot and prevents it from escaping. These dams also helped in storing most of the water on the plot, and have a storage capacity of approximately 1.5 to 1.9 million litres (Wilson, 2010: 4). To store more water to help manage the very unpredictable and highly variable rainfall in semi-arid Zvishavane, Mr Phiri Maseko embarked on the construction of canals in the 1990s. The canals are also a mechanism for spreading the water across the plot effectively making the plot a 'water plantation'.

Photograph 9: Fruit trees on Mr Phiri Maseko's plot

In addition to the dams, Wilson notes that in the 2000s, Mr Phiri Maseko:

> added about 180,000 litres of permanent storage capacity and 80,000 litres of temporary capacity to the existing approximately 1.5 million litres of the three ponds he has built and extended since 1973, and he has done so in ways that bring the water across the land even more effectively than his initial system of wells and canals were able to achieve. The water table has been transformed on his land and even below (from a speech by Wilson on Mr Phiri Maseko's Life Time Achievement Award on August 24th, 2010).

The 'water plantation' provides people in the villages of Ziyabangwa and Hlupo with clean drinking water. There are two wells and a borehole at his plot, rendering the Phiri Maseko family "spoilt for choice" in terms of availability of clean water in dry Zvishavane. The plot has made a difference in terms of water availability in this dryland part of the country to the wider community, as daily, during my fieldwork women, men and children come to fetch water (see Photograph 10**).**

Photograph 10: Sharing water - girls fetching water at one of the wells on Mr Phiri Maseko's plot

The rich resource of ground water has given birth to lush vegetation thus enabling his dream of creating a 'Garden of Eden' to reach fruition. It is home to many small creatures and rich avian diversity. Many birds roost in trees on the plot; his neighbours have named them the 'Phiri birds' (Wilson, 2010). Therein lies the paradox: these birds destroy some of Mr Phiri Maseko's crops as well as some of his neighbours' crops. In as much as Mr Phiri Maseko has been hailed as a champion of 'biospherical egalitarianism', he was not happy when his crops were consumed by the birds and other small animals. Biospherical egalitarianism is often defined as "the equal right to live and blossom", equality between nature and humanity (Naess, 1973). Consequently, he made scarecrows to try to scare the birds away. At times, his children scare the birds away. Mr Phiri Maseko's innovations as discussed this far, are a way of managing and mitigating risks and therefore building resilience. His innovations have helped him cope with a changing climatic environment. There is no doubt the water harvesting has helped him fight food insecurity.

Mixed farming

Ken Wilson writes, "The wide diversity of crops, livestock and other products provided him (Mr Phiri Maseko) with a steady and resilient income through the vicissitudes of economic and ecological crisis, cycle and change" (Wilson, 2010: 7). A diversity of crops ensures food security and maintains fertility on his plot to which he added compost and manure from his cattle pen.

To ensure that he gets a lot of manure from his cattle, Mr Phiri Maseko fed his livestock with crop residues when they are in the pen. This means that the more the cattle feed, the more the dung is produced. Remnants of crop residues decompose in the pen thereby increasing the levels of nutrients in the manure. Grass is also harvested on his plot and put in the cattle pen so that it decomposes and increases the quantity of manure he gets for the wide diversity of crops he grows. He teaches other smallholder farmers who visit him about the importance of organic agriculture.

To help control pests, Mr Phiri Maseko said he engaged in inter-cropping whereby he planted maize in the same field with cowpeas, pumpkins, sweet reeds and *nyovhi* (a vegetable that grows naturally in rural Zimbabwe and is said to be a pest repellent). He said that the practice of inter-cropping was handed down from generation to generation and is widespread among smallholder farmers. Matiza cited in Mangoma (2011: 14) says, "Pre-colonial cultivation was based on shifting cultivation with inter-cropping of crops". It is such local knowledge that Mr Phiri Maseko 'married' to outside knowledge to build his hybrid concept of agricultural practices. In addition to this practice, he practised crop rotation. For example, he rotated maize with groundnuts.

Mr Phiri Maseko also advised his visitors about the importance of diversifying their crops. He reminded them, "*Zvemono-mono siyanayi nazvo*" meaning that they should not practise mono cropping. Thus, in all his life as a farmer, monoculture was Mr Phiri Maseko's bete noire.

Growing many crops is a measure of agricultural risk management (Ziervogel, 2002). This adaptive measure is ideal in the wake of rainfall variability in semi-arid Zvishavane. Sifting through Mr Phiri Maseko's archives, I came across information compiled by Wilson in 1999 detailing the variety of crops he planted in that year. The crops included cotton, maize, sugar cane, elephant grass, reeds, kikuyu grass, *tsenza*, banana trees, beans, groundnuts, sweet potatoes and groundnuts. He also grew cassava and cowpeas (see Photograph 11). The seed he used to grow the diversity of crops is suitable for dryland Zvishavane.

Photograph 11: Mr Phiri Maseko holding a dish of cassava

Mr Phiri Maseko used seed he bought from seed houses (see Photograph 12) of him inspecting Pioneer Seed). This is contrary to assertions by Lancaster (2008) who writes: "Rather than using hybrid and genetically modified (GMO) seed, Phiri uses open-pollinated varieties to create superior seed stock as he collects, selects and plants seed grown in his own garden". I was sent by Mr Phiri Maseko to carry maize seed bought by his daughter from a seed house in Bindura (a small town about 75 km to the north of Harare). However, this is not to dispute that he used open-pollinated varieties as Lancaster asserts. He used his own seed to grow such crops as pumpkin, cowpeas, groundnuts, bambara nuts and many others. Mr Phiri Maseko did not limit himself to growing edible plants but non-edible ones as well that he sold to raise income.

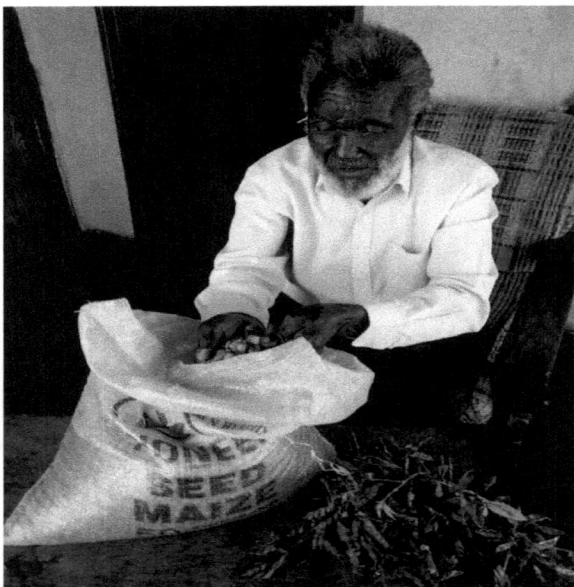

Photograph 12: Mr Phiri Maseko inspecting maize seed bought from Pioneer, one of the seed houses in Zimbabwe

In addition, he grew reeds on his plot that provided him with another income source (see also Wilson, 2010). The reeds are bought by basket weavers from other dry areas of Zvishavane. I interviewed a basket weaver who came to buy reeds at Mr Phiri Maseko's plot (see Photograph 13) and he said the following:

I want to thank Mr Phiri Maseko for harvesting water. Because of the water, a lot of reeds grow on his plot and therefore we can buy the reeds at a place, which is not far from our homes. The baskets I weave are mainly bought by workers at Mimosa Mine (in Zvishavane) and Unki (in Shurugwi) and business has been good. This has helped me to send my children to school (Interview with a basket weaver on 17 September 2012).

Photograph 13: A basket weaver transports reeds bought at Mr Phiri Maseko's plot with his donkey cart

Mr Phiri Maseko got many benefits from selling a variety of plants and seeds, which, like reeds, are not only a source of income and trade, but convey a message of the interdependence of water, soil, plants and people. This interdependence is captured in his use of metaphors, which the next section explores in detail.

Innovation and experimentation: What is the story?

The significance of Mr Phiri Maseko's story is that with predicted increase in rainfall variability, water management will be increasingly critical to smallholder livelihoods. Narain argues:

> ...it is clear that the management of water, and not scarcity of water, is the problem in many parts of the world...the solution, practised diversely in many different regions, lies in capturing rain in millions of storage systems...and to use it to recharge groundwater reserves for irrigation (Narain cited in Blackmore, 2007: 515).

Given that the area of study receives an average annual rainfall of about 546mm, rain-fed agriculture in Zvishavane does not yield much in terms of agricultural production hence the need for supplementing crops through water harvesting. With an anticipated drop of rainfall in southern Africa of below 500mm per annum, there is need to raise soil organic matter so that soil retains more rainwater. Alternatively, smallholder farmers could design structures that harvest and retain water such as Mr Phiri Maseko's 'immigration centre'. Innovations such as these have potential to make society build resilience and adapt to climate variability (see also Rodima-Taylor et al., 2012) given that based on climate change predictions, semi-arid environments would be unable to produce rain-fed crops such as maize or even 'orphan crops' such as sorghum and millet. However, the term 'orphan crops' has been contested. The discussion on the term will be left for another day.

This study has described Mr Phiri Maseko's techniques that have reduced his sensitivity to rainfall variability and built his adaptive capacity by his management of water and soil on his plot. For instance, I argue that he built resilience because of his vulnerability, i.e. exposure to climate variability in an area where the average amount of rainfall per annum is less than 600mm as shown in Graph1. From my interviews with Mr Phiri Maseko and his followers, I was appraised about intra-seasonal and inter-seasonal variability. They all agreed that rainfall was becoming more and more variable as compared to the past. One of his followers says he used to keep rainfall recordings that show evidence of increased variability. However, he has since misplaced the document. It is because of this increased variability that Mr Phiri Maseko and his adopters harvest water.

Data that I generated from my fieldwork suggests that one way that adaptive capacity has been built among smallholder farmers is developing innovations that are a synergy between local and external ideas. This study regards this coupling as *the* major characteristic of farmer innovation. In a world characterised by mobility it would likely be difficult to identify a wholly locally driven innovation. Smallholder farmer innovations are a product of the fusion of ideas from several sources. I therefore concur with Sanginga et al., (2009) who observe that for innovation to take place there must be inputs from various sources. This is evidenced by Mr Phiri Maseko's narrative as discussed in Chapter 1 that demonstrated how his innovations are a fascinating blend of modern technology and local knowledge that help boost food security.

This study highlights Mr Phiri Maseko's Malawian origins to show that his practices reflect regional practice. For example, I show that he grows cassava as a way of boosting his food security. Growing cassava is mainly a Malawian practice as suggested by Mr Phiri Maseko. He said that his father taught him how to grow cassava. Being of Malawian origin, suggests that his agricultural practices reflect a broader picture of southern African regional thought.

Mr Phiri Maseko was keenly aware that vulnerability in his variable environment was dependent on his hybrid conceptualisation of the interaction of local and external forms of knowledge in the socio-ecological system. The conviviality and hospitality that characterised his agricultural thinking and practices, extended therefore also to knowledges more broadly. For example, some of his innovations in terracing and water management had antecedents in the Nyanga Agricultural Complex in eastern Zimbabwe (Soper, 2000). However, Mr Phiri Maseko's terraces seem to be a local response to vulnerability. Another of his interventions, inter-cropping, was also practised by communities who lived in the Shashi-

Limpopo basin in pre-colonial times (Manyanga, 2000). Furthermore, the cultivation of vleis was a common practice among the Shona people of various times (Mangoma, 2011). In addition to these 'traditional practices', Mr Phiri Maseko also made use of modern ideas and techniques for example, using pipes in his fields for improved water management. Additionally, he uses different seed varieties with different drought resistance capacities. Therefore, he made use of both local knowledge and outside ideas to cope with climate variability.

The socio-ecological system is about interacting systems as Simonsen et al., explain, "...to investigate how these interacting systems of people and nature – or socio-ecological systems – can best be managed to ensure a sustainable and resilient supply of the essential ecosystem services on which humanity depends" (Simonsen et al., 2014: 3). Mr Phiri Maseko managed to reduce his sensitivity to variable rainfall by constructing structures that harvest water. He sought an "understanding of what is critical for building resilience and how an understanding of these factors can be applied" as Simonsen et al., (2014: 3) maintain. Mr Phiri Maseko's agricultural practices are innovative attempts of adapting to climate variability. A good example are the canals and small dams he excavated on his plot. They have potential to serve the Phiri Maseko family for decades to come. This case study demonstrates a smallholder farmer's capacity to reduce vulnerability in an era characterised by uncertainty.

This study suggests that Mr Phiri Maseko also demonstrated that adaptation is a continuous process of adjustment and change, both incremental and transformative, as argued by Bernier and Muizen-Dick (2014). The transfer of technology approach has failed leading to Hall (2009) asking, "Why are we still here?" This question brings to the fore the importance of revisiting dominant narratives on what constitute appropriate means of adaptation to changing climate.

Mr Phiri Maseko's agricultural practices are, therefore a fusion of incremental and transformative approaches to climate adaptation that can be situated within the typology of adaptation approaches discussed in Chapter 2.

His adaptation can be both autonomous and at a later stage, anticipatory. He started harvesting water before climate change became a topical issue on the developmental agenda due to the inherent rainfall variability in dry Zvishavane. He said:

> I started harvesting water after realising that rainfall days are few and far between here in Zvishavane. I made it a point that I would harvest water on the few occasions rain fell, a farmer worth his/her salt should never allow water to flow away because it is generally dry in agro-ecological region 4.

That he started preparing for rainfall variability by increasing, for example, the number of small dams from one to three suggests that his interventions were anticipatory. Due to increased climate variability, he started innovating in anticipation of more climate stress after studying intra-seasonal and inter-seasonal climate variability. Such anticipatory approaches (building dams and canals) are cost-effective in the end because all the smallholder does is to do maintenance work to keep systems functional.

His cost-effective innovations are closely related to the *chololo* pits by Mr Sungula from rural Tanzania that I discussed in Chapter 1. The common thread in these innovations is that they are inexpensive. Thus, innovations such as these are acceptable to many smallholder farmers and have the potential to help improve rural livelihoods.

Sustainable water harvesting and organic approaches to soil fertility have potential to improve rural livelihoods given that some of the smallholders cannot afford to buy synthetic fertiliser due to its high price. Mr Phiri Maseko's innovations illustrate how organic farming by means of adding compost and animal manure has increased his yields thereby consolidating his status as a *hurudza*. His farming practices reflect that an appreciation of the dynamics in one's environment is essential for smallholders to adapt to the effects of climate variability; but adaptation to changing climate should not only be seen through its negative effects. Crises should not only be perceived through their disillusioning effects on the ordinary people but should also be seen as unlocking latent innovativeness therein as evidenced by Mr Phiri Maseko's achievements.

Conclusion

At the heart of Mr Phiri Maseko's innovations is the realisation that rainfall in Zvishavane is erratic and unreliable. The techniques of harvesting and channelling water described in this chapter have ensured continuous water supply for his crops over many years, and from the range of accounts in conversation, and in written sources, as well as in the many awards he received, it is evident that he managed to raise the water table on his plot in a very dry area. His innovations helped transform his plot into a 'Garden of Eden' and this blend of fighting attitude and technical aptitude that characterise him as a *hurudza*. His infrastructural innovations that have been shown in detail are the physical manifestation of how he thinks about soil and water, there are many kinds of structures, but they reflect a consistent set of principles as set out in Chapter 3. They are grounded on a philosophy of soil and relationality that draws on regional southern African thought and practice.

Therefore, the next chapter argues that the combination of innovations grounded in regional philosophy of soil/water and relationality are indeed very hard work to create, if not to maintain. The question from the point of view of agricultural policy makers who are trying to conceptualise interventions appropriate for climate change, is whether these interventions would be carried out, given their hard work. Chapter 4 explores the uptake of these interventions, arguing that they have indeed been successfully taken up by a completely wide range of farmers in the region, despite the challenges facing ZWP. Mr Phiri Maseko's practices have traction and people in a wide area around Zvishavane have been willing to invest in them.

Chapter 4

Rhyming with an audience

Introduction

"Why should we deepen our contours?" asked a despairing teenage boy to his father, a smallholder farmer in the Mapirimira Ward of Zvishavane. The teenage boy seemed oblivious to the fact that a local smallholder farmer, Mr Phiri Maseko, had deepened his contours to good effect. However, the teenage boy's sentiments amply sum up the inescapable sense of a dystopic future in semi-arid Zvishavane as smallholder farmers lurch from one drought to the other. It takes a lot of hard work and dedication to implement innovative agricultural practices such as deepened contours that help in boosting rural livelihoods. Whether or not people innovate using practices that are labour-intensive as Mr Phiri Maseko's deepened contours, is a crucial question for policy-makers and agricultural advisors.

This chapter documents the work of smallholder farmers who have implemented Mr Phiri Maseko's water harvesting techniques and the improvements of their livelihoods. I examine the barriers that hinder uptake of innovations by smallholders and enablers that enhance uptake. The farmers (adopters) who have copied his techniques constitute the audience that rhymes with him. It is inaccurate to suggest that Phiri Maseko adopters' success is to their duplicative efforts, because they too have tinkered and innovated creatively in their water harvesting innovations. In other words, rhyming with him implies that the audience is inspired to do more creatively than just merely copying the Phiri Maseko innovations to be able to adapt to their unique local environmental dynamics. The capacity to adapt to one's environment (rhyming with Mr Phiri Maseko) is enhanced by visiting and learning from him at his plot and subsequently followed up through implementation at home. These are innovators characterised by "ceaseless curiosity" as demonstrated by Mr Phiri Maseko (their mentor) at his home. His plot has become a cosmopolitan crossroads for smallholders, academics, researchers, students, NGO officials and agricultural extension workers from around the world.

In this chapter, I unpack the term 'rhyming' with an audience. I derive this term from Mr Phiri Maseko who said that he "rhymed with nature", meaning that he engaged in sustainable ecological practices as evidenced by his marriage of water and soil. He also said of a farmer in his village who has implemented his farming practices that he "rhymes with me". Rhyming with his audience conveys more than merely learning from his methods of water and soil management.

Central to Mr Phiri Maseko's innovations is the slogan *kohwa mvura, kohwa pakuru* (harvest water, harvest a bumper crop). Ideally, this means that if one harvests water then one builds ecological resilience and one realises higher crop yields because of water availability. The slogan *kohwa mvura kohwa pakuru* (see Photograph 1 of Mr Phiri Maseko wearing his t-shirt with these words) was popularised by Africare an NGO operating in Zvishavane. While the *kohwa mvura* concept will be discussed later in this chapter, it is important to note that the genesis of this slogan can be attributed to Mr Phiri Maseko whose relentless work ethic has been instilled in his adopters.

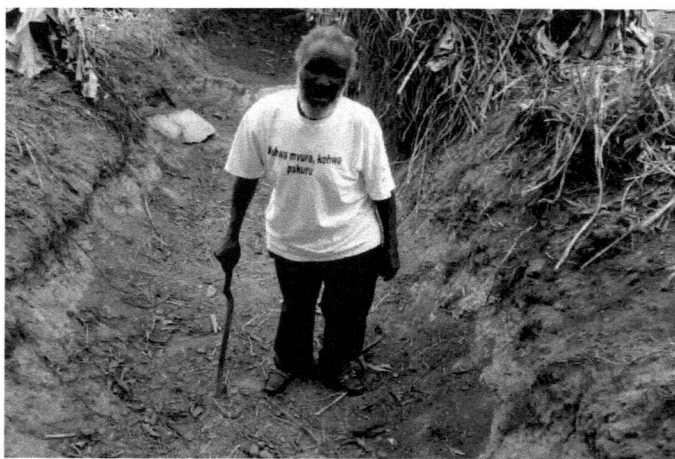

Photograph 1: Mr Phiri Maseko wearing his 'kohwa mvura kohwa pakuru' t-shirt stands in a deepened contour at his plot

Mr Phiri Maseko offered tangible adaptive climate responses through his innovations. He motivated other farmers - his motley band of cohorts - who are vulnerable to rainfall variability in southern Zimbabwe. Innovative smallholder farmers such as Mr Phiri Maseko helped other farmers navigate barriers through his water harvesting techniques. After all, smallholder farmers closely work with one another in rural Zvishavane that helps to enhance their adaptive capacity. Many farmers visited Mr Phiri Maseko to learn about his innovative adaptation methods because they understand the imperative to harvest water. This demonstrates he rhymed with other smallholder farmers and helped them fight food insecurity. This chapter discusses the smallholder farmers who learnt from Mr Phiri Maseko; these are the farmers who rhymed with him. His conceptualisation of water and soil resonated with other smallholder farmers. Mr Phiri Maseko rhymed with his audience of smallholder farmers because he became an icon, an award-winning smallholder farmer who provided inspiration and moments of excitement.

"Adaptation by ribbon cutting?" NGO interventions in Zvishavane

Introduction

Agricultural NGOs are increasingly active in rural Zimbabwe because of government's inability to improve smallholders' livelihoods. The country's failing economy in part due to the hyperinflation from 2007 to 2008, and the reality that Zimbabwe continues to experience economic decline with job losses and factory closures resulting in an urban-rural migration have exacerbated this situation. To boost their livelihoods, the 'new arrivals' engage in smallholder farming.

I wish to stress that this study will mainly use the phrase "adaptation by ribbon cutting" to assess the role of ZWP in rural Zvishavane and its neighbouring environs. The study will assess peripherally the other NGOs operating in the same environment as ZWP as most of them were reluctant to allow me an opportunity to research on their activities. The role of NGOs in rural Zimbabwe is a volatile issue (Mawere and Mabeza, 2015). The government suspects that NGOs support the opposition and this has created an uneasy relationship between them. From my fieldwork in rural Zimbabwe, I encountered several NGOs doing work to assist smallholder farmers adapt to changing climate including the Zvishavane Water Project (ZWP), an indigenous NGO founded by Mr Phiri Maseko.

This chapter seeks to understand NGOs working with smallholder farmers. I argue that some of the NGOs have launched projects that have genuinely assisted the rural communities to cope with a changing climatic environment; the additional reality is that many of them collapsed once donor funds had been withdrawn. It is worth noting that Mangoma (2011) cautions against the tendency to view collapsed NGO projects as total failures. In research on a wetland development project in semi-arid rural Zimbabwe, the New Gato Water Project (NGWP), Mangoma (2011: 241) argues that despite the disintegration of the NGWP, the local people benefited by implementing what they had learnt from the project. Mangoma adds that technology transfer is often overlooked by critics of NGO interventions. Technology transfer was certainly evident in my fieldwork in that smallholder farmers were implementing ideas they learnt from some NGOs. For instance, smallholder farmers in Zvishavane and its surrounding areas of Shurugwi and Chivi are digging basins to plant crops and using mulch to preserve moisture to good effect as they had directly learned from NGOs that are promoting CA. Nonetheless, it must be noted that some smallholders have not adopted NGO technologies and the following section explores possible explanations for this lack of uptake.

NGO interventions for managing climate variability in Zvishavane and its environs

Zimbabwe's rural areas are home to interventions initiated by NGOs. Some of the NGOs' interventions are in the form of inputs (inorganic fertiliser, seed). NGOs often run competitions that offer grand prizes, for example, cows (see Photograph 2) in a bid to motivate smallholders to adopt their agricultural practices. Uptake of innovations of this nature is rapid because access of free inputs is attractive to smallholders. However, the duration of most of these projects is short to the extent that their successes are not easily noticeable. Most of the smallholders eventually withdraw after realising that inputs are mainly reserved for a few, the so-called lead farmers.

Photograph 2: A smallholder farmer shows a cow he won in a Conservation Agriculture (CA) competition

The poorer smallholders have little, if any, access to inputs. Without financial resources to buy agricultural inputs such as fertiliser, smallholders are unable to implement many NGO interventions even though some of these have shown promise.

There are at least eight agricultural NGOs involved in implementing projects in Zvishavane including ZWP. Their details are shown in Table 1.

NGO*	TYPE OF INTERVENTION	OPERATIONAL WARDS	NUMBER OF HOUSEHOLDS	DURATION
NGO 1	CA** Basins, infiltration pits, tied ridges ripping into mulch, fanya juu, standard contours	2,9,6,11,12,13,17, and 18	120 hh*** per ward	2 years
NGO 2	CA basins	1,2,9,4 and eight	1700 hh	On going
NGO 3	CA basins	15,11,16 and 5	400hh	On going
NGO 4	CA basins	14 – 18	1436 hh	5 years
NGO 6	CA basins	1,2,9,4 and 8	Number not given	3years
NGO 7	Goat project	14-18	Number not given	On going
NGO 8	Supplementary feeding	6	Number not given	16 years
ZWP	Dams, wells	All wards	Number not given	8 years

Table 1: Interventions by NGOs in Zvishavane (Source: Interview with an AGRITEX official, 11 November 2012, Zvishavane)

* Due to ethical considerations, I decided not to name the NGOs doing interventions in Table 12, as some of them were not comfortable with their names being published. I mention the names of NGOs that were not apprehensive about their names being published. However, I could not disguise ZWP, as it is an integral part of Mr Phiri Maseko's story.

** Conservation Agriculture

*** - households

A few of these NGOs are engaged in implementing CA. CA is known in the rural areas as *dhiga udye* (dig and eat) while 'cheeky' villagers have termed it *dhiga ufe* (dig and die) in apparent reference to the fact that the process is so labour intensive that it might send those who practise it to an early grave. However, not to be outdone, some NGO officials with a keen sense of humour have hit back at their detractors by rebranding the *dhiga udye* concept as *dhiga ufe nekuguta* (dig and 'die' of satisfaction from eating too much food). At one of the meetings to introduce CA in neighbouring Shurugwi District, a village head promptly stood up while an NGO field officer was in the middle of his speech and said:

> *Moda kutikovedza,* (You want to kill us). Most of the people gathered here are ailing and so how are they going to cope with such a demanding project like *dhiga ufe*. My people can get involved in this project; I will not stop anyone from participating. Most of the people here are also too old and can no longer cope with such demanding tasks. I am gone.

CA involves digging basins where the seed is planted. The farmers involved in CA harvest grass for mulching. Inputs for CA are only given to the lead farmer in a group of about ten farmers by the NGOs involved. In turn, the lead farmer influences other farmers to join his/her group. The lead farmer only receives inputs on condition that he/she has a group of about nine farmers, the so-called mentored farmers. During my fieldwork, I observed that the rest of the farmers do not harvest much because of lack of inputs. It is worth stating that both water harvesting and CA are a lot of work, but Mr Phiri Maseko's system is not so dependent on purchased inputs. Thus, CA is not sustainable in the long run. In her study of "people based conservation" of the Madikwe Game Reserve in South Africa, Bologna (2008) makes a similar observation:

> There was one Mafisa-led project, the internship programme that aimed to train eight young people in lodge management, tour guiding and game ranging. But, again, the project was characterised by the immediacy of its goals - to transfer the necessary skills to the interns within a very limited period, one defined by the funding available. The

project was thus not sustainable over the longer term - most obviously because it was a training course and therefore entirely dependent on the NGO with its intermittent, exogenously controlled cycles of donor funding (Bologna, 2008: 194).

Yet some NGO projects have potential for sustainability. To its credit, NGO 1 has been promoting the cultivation of small grains such as sorghum and the programme was running from 2010 to 2012. The response was encouraging in the sense that about 300 farmers in Mapirimira Ward grew small grains. One of Mr Phiri Maseko's adopters, Mr Mawara who lives in Mazvihwa area of Zvishavane, harvested roughly 42 bags of sorghum in the 2011/2012 farming season. Mr Mawara says deepening contours alone is not enough without seed. He appreciates the role played by NGO 1 in distributing seed of small grains to smallholders.

NGO 2 is involved in the dissemination of information on climate change and HIV/AIDS. It also supports savings clubs, poultry, and gardening projects in Ward 6. In Hlupo village, NGO 2 has initiated a poultry project for the smallholder farmers. Mrs Phiri Maseko coordinates this project and a youth named John Sibanda. Besides poultry farming, the members also weave baskets and mats, which are sold in Zvishavane. NGO 2 brings smallholder farmers from neighbouring countries to their counterparts in Zvishavane. One such event was organised in Hlupo village, where the hosts displayed some of their products that were bought by visiting smallholders from Mozambique, Malawi, Lesotho and Zambia.

NGO 8 focused on provision of supplementary feeding programmes to students in Ward 6 schools from 1994 to 2010. They also helped to construct a dam near Maketo village in Mapirimira Ward under the food-for-work programme. NGO 8 also initiated a garden project near the dam and had it fenced. The organisation also connected pipes from the dam to the garden that have since been vandalised. Other NGOs such as ZWP implemented revolving goat projects.

Based on the activities of NGOs in Zvishavane mentioned above, this study suggests that generally they are contributing towards boosting food security. Thus, some of their interventions do not amount to 'adaptation by ribbon cutting'. The section below gives an account of the activities of the ZWP because Mr Phiri Maseko had a long history with the organisation. It is out of this association that he spread his ideas in Zvishavane and neighbouring districts.

Adapting to rainfall variability: Zvishavane Water Project

"An attempt to do something beyond the usual-usual of development" is one of the major objectives of ZWP (Wilson, 2010), a registered local NGO that was started in 1987 by Mr Phiri Maseko and emerged out of a collection of his experiences as a pioneer of sustainable wetland farming techniques in rural Zvishavane according to Charles Hungwe, its first administrator. Mr Phiri Maseko worked as a field co-ordinator from 1987 to 1996 when he retired, although he has remained in contact with the organisation on an advisory basis. ZWP is one of the lasting legacies of Mr Phiri Maseko to Zvishavane and Chivi Districts of Zimbabwe as evidenced by its impact on smallholder farmers who are now practising his water harvesting techniques. However, do its interventions amount to "adaptation by ribbon cutting?"

ZWP rose out of the efforts of Ken Wilson and Ian Scoones together with Mr Phiri Maseko. Ken Wilson was doing his doctoral fieldwork in Mazvihwa and he invited Mr Phiri Maseko to come and assist other smallholder farmers with water harvesting techniques. Ken Wilson linked Mr Phiri Maseko to Phillip Cole of the World Development Programme (WDP) in the UK to raise funds to spread his ideas in Zvishavane. Cole invited Mr Phiri Maseko to the UK in June 1987. WDP agreed to fund the operations of ZWP and the money was to be channelled through Oxfam GB in Harare. That same year ZWP started operations at rented premises in Zvishavane town.

ZWP is explicit about its purpose as captured in its organisational logo that "Water is life" (see Fig. 1). Part of ZWP's mission statement reads:

A Zimbabwean NGO that facilitates and promotes community based initiatives in water harvesting, sustainable agriculture, environmental management processes and capacity building interventions designed to alleviate poverty and sustain food and income security in southern Zimbabwe (ZWP, undated: 2).

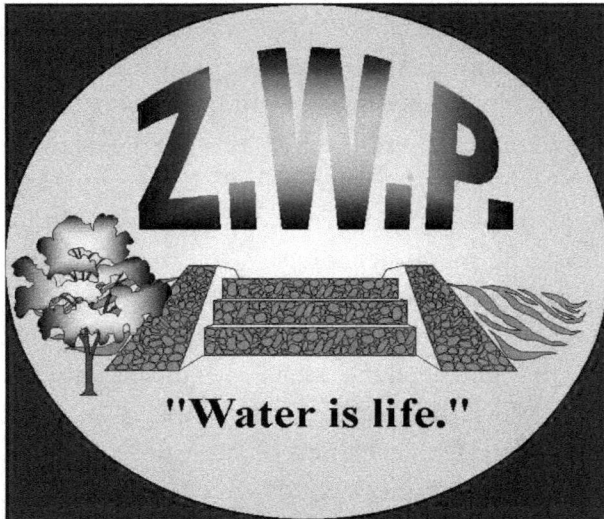

Figure 1: ZWP logo (Source: ZWP, 2012)

A ZWP article of November 1987 states that its interventions in Zvishavane and Chivi mainly promoted water harvesting as follows:
- *Ruware* run-off tanks – brick and concrete – brick and concrete lined tanks 2m deep and 20m in diameter designed to catch rainwater from the large smooth granite rocks that are characteristic of the area. The water harvested is used for small gardens and nursery plots
- Community gardens – the gardens had concrete–lined channels laid to allow irrigation of small garden plots
- Rainwater harvesting from roofs. The tanks were circular and concrete-lined (3m high and 6m in diameter) the tanks collected water from roofs of buildings such as schools. The tanks at schools are meant to help in providing water for school gardens

- Improvements were made to existing wells by deepening and lining them, adding windlasses to facilitate drawing of water and partially covering them to prevent contamination
- In the plains many people relied on *mifuku* (holes in the sand of dry streams or river beds) the holes are vulnerable to contamination and drawing water directly from the sand is tedious. A method was devised of lining the *mifuku* with covered concrete rings, which greatly improved the supply rate and quality of the water albeit this method does not produce very safe water (ZWP, November 1987: 2).

ZWP's vision was to operate in six districts of the country, whilst it currently operates in four districts Zvishavane and Mberengwa in the Midlands Province, and Chivi and Mwenezi Districts in Masvingo Province. ZWP is operating in four wards in Chivi; these are 11, 12, 15 and 16. Mr Phiri Maseko suggested that ZWP's mandate to sink wells was meant to reduce the incidence of water borne diseases and ensure a balanced diet through the provision of nutritional gardens.

The nutritional gardens served to spread the ideas of Mr Phiri Maseko. Ian Scoones concurs in his citation on the nomination of Mr Phiri Maseko for the 2004/2005 King Baudouin International Development Prize:

> From his initial work at his own home, this spread across Zvishavane district through the establishment of Zvishavane Water Project. The small amount of seed money provided by Oxfam GB and an EU small grants scheme allowed the extension of activities across Zvishavane into neighbouring Chivi district. This attracted the attention of others – both ordinary farmers and organisations working in other places. Soon Mr Phiri Maseko was being invited all over the country to share his ideas. And his home in Zvishavane communal area became a place visited almost on a daily basis. Farmers' groups from Zimbabwe and post-1994 from South Africa visited regularly, as did extension workers, researchers, policymakers and others. All left inspired, and the effects can be seen in the enterprising work carried out by farmers – infiltration pits, rock catchments, wells, wetland

gardens and so on soon spread across the countryside. This was not a 'big project' approach. The ZWP never had a large amount of external support; this was locally led development, responding to real needs and ambitions of ordinary farmers, inspired by a remarkable man and his dedicated team (Scoones, 2004).

The first few projects set up by ZWP were for small dams and soil and water conservation. The pioneer donor partners included World Development Programme through Oxfam GB and Ireland, and Environment and Development Activities - Zimbabwe (ENDA - Zimbabwe). The financial and human resources at ZWP's disposal were limited but increased steadily. With the available resources, ZWP did activities such as holding workshops. The workshops were meant to equip Mr Phiri Maseko's audience with technical skills on water harvesting in semi-arid southern Zimbabwe. One such workshop was held in 1993 at Msipani's Maketo Community Hall in Mapirimira Ward. The two-day training workshop was facilitated by Mr Phiri Maseko. The annual ZWP report of 1993 stated that smallholder farmers who participated were drawn from Zvishavane and other districts in Zimbabwe such as Mberengwa, Chirumanzu and Chivi. All in all, about 200 smallholder farmers were present at the workshop. The theme of the workshop was "Water conservation using local technical knowledge". Participants were taught about constructing infiltration pits and earth dams. The participants learnt skills they used to construct wells and dams in their villages with the assistance of ZWP.

During Mr Phiri Maseko's stint at ZWP, about 80 wells were sunk and 35 small dams constructed, mainly in the Mazvihwa, Mapanzure, Mafala, Wedza and Masunda areas of rural Zvishavane. He implemented his water harvesting techniques primarily in Mazvihwa because that was where he was based during his engagement with ZWP, in the 1980s. Mr Phiri Maseko with three other 'lecturers' tutored at the 'University of Mototi' where Ken Wilson and Ian Scoones, among others, were 'students'. The informal learning environment at the 'University of Mototi' was replete with 'students' and 'lecturers' who were extremely knowledgeable and skilled. This informal learning environment resonated with participatory and collaborative research, a theme that is of importance according to Veteto and Crane (2014). Collaborative research at the 'University of Mototi' involved Mr Phiri Maseko and other farmers together with researchers from abroad such as Wilson and Scoones. Crane says, unpacking the "back-to" part of "farmer-back-to-farmer" means acknowledging researchers as stakeholders in the process, just as much as farmers are (Crane, 2014: 49)

'The University of Mototi' was instrumental in helping other smallholder farmers rhyme with Mr Phiri Maseko. 'Established' in the mid-1980s, the 'University of Mototi' was the precursor of Mr Phiri Maseko's outreach work with ZWP. Ian Scoones (Wilson, 2010) says the 'University of Mototi campus' was located at Mr Cephas Mukamuri's home at a rural business centre in Mazvihwa and close to the Runde River (see Photograph 3). Scoones (2010) says that the 'tutors' at the 'University of Mototi' were world class. These included Mr Cephas Mukamuri as the 'vice chancellor', 'Dr' Mathou Chakavanda, 'Dr' Tangwena and 'Professor' Zephaniah Phiri Maseko. Mr Cephas Mukamuri was an educationist of repute in the Mazvihwa area. His son Billy 'graduated' at the 'University of Mototi' together with Ken Wilson and Ian Scoones. The 'graduates' ran extensive workshops with smallholder farmers on the Phiri Maseko water harvesting skills. The 'University of Mototi' 'alumni' still engage the smallholders in Zvishavane.

Ken Wilson (just like Ian Scoones) makes a 'pilgrimage' to Zvishavane almost every year to 'give back' to the smallholders (Mr Phiri Maseko's audience) and does a lot of extension work that includes giving technical expertise on small dam construction thus standing in for the government extension workers who are unable to fulfil their mandate. Based on a discussion I held with Ken Wilson at the now defunct 'University of Mototi' on 14 January 2013, there could be as many as 1000 farmers in Zvishavane alone implementing Mr Phiri Maseko's water harvesting techniques. This he said can be attributed to the sterling work done by the 'lectures' at the 'University of Mototi'. Smallholder farmers who 'graduated' from the 'University of Mototi' include Mr Banda and Mr Mawara, whose innovative agricultural practices are documented in this chapter.

Photograph 3: At the now defunct 'University of Mototi'

The 'lecturers' at the 'University of Mototi' impressed on their 'students' to engage in preserving trees and harvesting water. At a community meeting on 13 September 1987 at Nyevedzana, in the Gudo Village of Mazvihwa, two of the 'lecturers,' 'Dr' Mathou Chakavanda and 'Prof' Phiri Maseko, taught the villagers about conservation of trees and water projects. Below is an (unedited) extract from Mr Phiri Maseko's diary about what transpired at the meeting:

13 September 1987: I went to the Marozve HQ to take Mr Mathou [Chakavanda, my colleague in the 'University of Mototi' research team and the lead on indigenous trees] to Gudo as there was a tree meeting on this day. We got there at 9:20am and the meeting started at 10am. The VIDCO [Village Development Committee] secretary... asked Mr

Zezi to give a word of prayer and then he gave the meeting to Mr Mathou. The meeting was attended by 25 people.

Mr Mathou introduced me to these people and then told them we were working under the Research, and that Phiri Maseko is working on water projects, wells and dams. Mathou said we get all the information of your need from you. Therefore, we want to talk about your VIDCO today over your grazing area. Do you have trees there? Yes, they said. He asked, which [additional] trees could you want in your grazing? They said they would need Mutobwe, Musumha and many others. Mr Mathou said, "Do you know that you are the people using these trees? So it is you again who are going to replant them in your grazing.

Mr Mathou asked which are the (traditional) laws that make you keep your trees? [They answered:] "No one is allowed to just cut a tree, and you are not allowed to cut two or three trees near each other. You can cut trees at about 50m apart".

Mr Mathou also gave me time talk. I said I am also working under the research team but my work is on water projects such as wells, dams and also gardens and wetlands. I said to have a good garden is when you have water. To have clean people is when you have water. To plant nice orchards is when you have water. So for people to have good water for drinking is when you shall have sunk that well. So I have all means to show you how you can dig and line the wells and also when you collect the Z$20 we give you cement to line your well.

One woman said "Gudo Dam needs to be scooped all the mud; could you Phiri do that?" I said "You people have to do that work" (Extract from Mr Phiri Maseko's diary September 13, 1987).

Drawing on the above outreach meeting, it appears that Mr Phiri Maseko had the unenviable task of trying to impart the spirit of self-reliance on fellow smallholder farmers who seem to have been affected by a dependency syndrome. Some of the farmers get handouts in the form of food and seed from NGOs.

ZWP constructed water harvesting tanks in rural Zvishavane at schools (see Photograph 44) that include Mbilashaba, Rusvinge, Msipani, Shiku, Gudo, Ndinaneni, Korogwe and Gwemombe Dip

125

(Zvishavane Water Project, 1992). I visited some of these schools to assess the efficacy of these tanks. Most of the tanks are in a state of disrepair due to the impacts of the economic crisis in Zimbabwe. Maintenance of the tanks was no longer a priority as the small amounts of money available in schools was spent on buying books and other necessities. The story behind the construction of tanks at Gwemombe Dip School (see Photograph 4) bears testimony to Mr Phiri Maseko's charm. It is also testimony to his ability to empathise. One cold night as we warmed ourselves around a fire, Mr Phiri Maseko who had been dozing, suddenly became animated to life and recounted:

Oh, there is this story about the tanks at Gwemombe Dip School I forgot to tell you. One day in the 1990s, I visited the school and found teachers in a sombre mood. I enquired from the headmaster why teachers appeared to be very unhappy. The headmaster said that the teachers were unhappy with the dire water situation at the school. The wells had dried up and teachers were fetching water at a borehole that was more than 5 km away. Thus, about 90% of the teachers had requested to transfer to other schools. I sought permission from the headmaster to address the teachers and it was granted. I impressed upon the teachers that a solution could be found and that he would ask the ZWP to construct tanks that harvest water that flowed down the rock outcrop at Gwemombe. I promised the teachers that the water problem could be solved and they agreed not to transfer. We immediately started the work of constructing the tanks. After finishing the construction, a lot of water was harvested during the rainy season. The school started to do market gardening and teachers were allocated small portions in the garden to grow their own vegetables. I used to meet some of the teachers in Zvishavane town and they were very thankful to me about the work ZWP had done for their school. Zephaniah *bakithi!*

Photograph 4: Water harvesting tanks at Gwemombe Dip School

ZWP does not only promote water harvesting, it also encourages farmers to cultivate 'orphan' crops such as millet, sorghum and rapoko as a viable option to adapt to rainfall variability in dry land Zvishavane. Witoshynsky notes that Mr Phiri Maseko had addressed this:

> At ZWP we are discouraging maize cultivation. Okay, we know maize is an easy crop to grow, easy to harvest, easy to cook. But the problem is that it is unable to stand against drought so we encourage farmers really to look into our traditional crops such as our millets and sorghum (Witoshynsky, 2000: 49).

The interviews I conducted revealed that 'orphan' crops provided the staple diet for some farmers who live in southern Zimbabwe. 'Orphan grains' include rapoko, millet and sorghum. The name orphan suggests that the crops have been either neglected by science or underutilised considering their potential (Pereira, 2014). 'Orphan crops' are drought resistant and suitable to dry land Zvishavane. However, during my fieldwork I realised that maize remains the crop grown by most the smallholder farmers despite persistent drought. The reason might be that growing 'orphan crops' is very labour intensive. Moreover, the 'orphan crops' are easily attacked by pests. Because of their small grains, birds are the bane of the 'orphan crop' farmer. Regardless of these hurdles, 'orphan crops' offer an innovative strategy for addressing smallholders' vulnerability to drought. There has been an increase of smallholders who grow 'orphan crops' in Zvishavane perhaps because of the activities of ZWP and other NGOs that are also promoting 'orphan crops' as a way of adapting to rainfall variability. The smallholder farmers such as Mr Mawara preserve their grain in the family granaries for five years (see Photograph 5). Mr Mawara said that he cherishes the time he spent with Mr Phiri Maseko before his retirement from ZWP.

Photograph 5: Inside Mr Mawara's granary full of rapoko

On his retirement, Mr Phiri Maseko donated a piece of land to ZWP for use as a demonstration garden for smallholder farmers under its auspices. However, due to ZWP scaling-down operations, the piece of land is now used by one of the sons of Mr Phiri Maseko to grow vegetables. Mr Phiri Maseko's son waters his crops using water from a well sunk due to his father's initiative.

Some of the wells sunk by Mr Phiri Maseko are still functional. Others are now disused. When I visited Mazvihwa in the company of Mr Phiri Maseko, he looked dejected to discover that one of the wells is now disused. The villagers said that they no longer used the well because the pump and lid had been vandalised and so the water posed a health hazard. Villagers now walk about three kilometres to fetch water. In contrast, the well in the village of Batakati is still functional. This can be attributed to the work done by the village head that instilled a sense of ownership into the villagers.

Keeping the projects functional has helped to disseminate Mr Phiri Maseko's ideas. The tanks constructed at schools helped to spread his water harvesting techniques among students, some of whom were to implement his farming practices. In implementing these projects, ZWP learnt by trial and error. After its formative years, it occurred to ZWP that if not properly implemented, projects such as dam construction would not give the smallholders maximum benefits. Mr Phiri Maseko and his co-workers decided to implement a 'code of ethics', (a set of guidelines for dam construction) that rural communities were to adhere to. Some of the guidelines they agreed upon included that rural communities were supposed to construct sand traps on the dam site area and that a dam site was required to have about fifteen sand traps (ZWP, 1995).

The more enterprising smallholders implemented these water-harvesting techniques such as the construction of sand traps at their homes. This is a milestone on the part of ZWP as its interventions continue to reverberate in rural Zvishavane as evidenced by smallholder farmers who use skills they acquired through ZWP to construct dams at their own homesteads. They are utilising water from the dams to do market gardening. These farmers (Mr Phiri Maseko's adapters) have realised the importance of self-reliance in order to secure livelihoods in changing climate and they are discussed later in the chapter.

ZWP's involvement with communities underscored the need for self-reliance, in a context where over-reliance on donors slows down creativity and willingness to innovative among the smallholders. Organisations such as ZWP are taking the lead in teaching smallholders to be self-reliant. The smallholders were involved in the projects implemented by ZWP as evidenced by the work they did. Instead of ZWP supplying them with all the building materials, the villagers dug wells, gathered and crushed stones, constructed sand traps, etc. Teaching the villagers how to construct wells, for example, was a way of teaching them to be self-reliant.

After 1997, in the post-Phiri Maseko era, ZWP's interventions shifted from projects that were mainly premised on water harvesting. This perhaps was a result of the requirements of donors funding the projects. For example, in 2011 ZWP launched a programme in Chivi and its purpose was to "improve household food security and reduce water borne diseases for 1750 selected households through a combination of broad interventions that include agriculture support, water and sanitation and social protection in Chivi by August 2012" (ZWP, 2012). Smallholders were taught such activities as borehole rehabilitation (see Photograph 51). ZWP also embarked on a small livestock distribution and about 750 households in Chivi's Wards 11, 12, 15 and 16 received goats, chickens, turkeys and sheep (see Photograph 6) (ZWP, 2012: 15). ZWP did capacity building and trained households on small livestock management. ZWP (2012) says that the villagers are now realising benefits through small livestock.

In 2012 ZWP launched a project that sought to assist "women farmers in Zvishavane and Chivi in making economic use of their readily available natural resources and increase their disposable income" (ZWP 2011: 1). The natural resources the project utilised are amarula, mobola plum and baobab. The project hopes to promote sustainable harvesting techniques and promote environmental care so that future generations can benefit (ZWP, 2012: 1). The amarula, baobab and mobola plum fruits are famed for their nutritional value.

The oil extracted from amarula and mobola plum and powder from baobab are sold to local and international markets. During a discussion with the women involved in this project in October 2012, I learnt that the women had been busy marketing their products at international trade exhibition events such as the Zimbabwe International Trade Fair and Harare Show with encouraging outcomes. I also sampled a drink from baobab powder the women make, known in Shona as *mahewu* that they are marketing.

Photograph 6: Villagers involved in borehole rehabilitation (Source: ZWP, 2012)

The account above indicates the diversity of activities by ZWP since Mr Phiri Maseko's retirement in 1996. It became more and more reliant on donor funding. Recently, ZWP has scaled down its operations as evidenced by the skeletal staff remaining at its offices partly because many of its donors have withdrawn due to differences with the Zimbabwean government, and consequently most of the senior staff has left the organisation due to ZWP's inability to continue paying them salaries. This brings to the fore the question about NGOs not staying the distance. It seems NGOs that are reliant on foreign donors normally collapse after the donor withdraws. Chapter 1 provided an example of the *Ngwarati* Wetland-Tillage System and how it collapsed after the Australian government stopped funding it. Despite all the uncertainty surrounding ZWP, water-harvesting techniques from its formative years remain a beacon of hope towards successful adaptation to climate variability for the Phiri Maseko adopters in southern Zimbabwe. Therefore, based on the above account, ZWP's interventions do not amount to "adaptation by ribbon cutting" because it genuinely assisted smallholders. The next section explores the experiences of farmers who have replicated Mr Phiri Maseko's innovations.

The "pro-truthers?" Adopters, adopters' adopters, adapters and "Josephses of Arimathea"

Innovations work if they rhyme with an audience. Zvishavane is home to such innovations by rural farmers, the "pro-truthers" (taking a cure from "The Economist") those who have defied the doomsday approach by means of their ingenuity. The Economist (September 2016) views "pro-truthers" as individuals who rebut the language of the establishment. The innovative techniques by the "pro-truthers" fly in the face of the "end is nigh" mantra of the prophets of doom. Smallholders have adapted to climate variability over the years, thus amply demonstrating that human ingenuity can play an important role towards the continued existence of humans in the face of adversity. Mr Phiri Maseko casts himself as the leading "pro-truther".

Mr Phiri Maseko implemented water and soil management techniques, which rhyme with other smallholder farmers. These techniques have ensured adequate food for the Phiri Maseko family. His ideas have circulated to may smallholders, who are implementing his water harvesting ideas, both the adopters and those who duplicate from Mr Phiri Maseko's adopters the adopters' adopters. Some of the adopters' adopters do not even know about Mr Phiri Maseko. Mr Magogomere, for example, one of Mr Phiri Maseko's adopters, says that some adopters copy from Mr Phiri Maseko and "disappear"; they do not go back to their mentor for more technical advice. Mr Magogomere uses the analogy of Jesus Christ's disciple, Joseph of Arimathea, who he says as a disciple of Christ, that he is 'invisible' and only presents himself after the death of Christ. He calls these adopters "Josephses of Arimathea". Mr Magogomere says:

> There are some of Mr Phiri Maseko's adopters who I call 'Josephses of Arimathea' because they remain invisible. They do not visit their master to learn more about his agricultural techniques and potentially might only visit after Mr Phiri Maseko's demise and claim

to be his followers. These compare very well to a follower of Jesus Christ who requested for the body of Christ and took it to a tomb, Joseph of Arimathea. Joseph of Arimathea only became visible after the death of Christ (Interview with Mr Magogomere, 7 December 2012).

Mr Phiri Maseko and his followers such as Mr Mawara, Mr Banda and Mr Batakati are some of the farmers involved in water harvesting in rural Zvishavane. Most these farmers are in the Mazvihwa area. This is where Mr Phiri Maseko earned himself the nickname *VaMamvura* (Waterman) because of his water harvesting activities. Perhaps his nickname can be attributed to his water harvesting activities in Mazvihwa during his years as a fieldworker with ZWP. He sunk many wells in the area and taught farmers water-harvesting techniques.

However, it is discouraging to note that a very negligible number of people in his own village are implementing Mr Phiri Maseko's innovations. Several perspectives may explain this scenario. Some villagers despised Mr Phiri Maseko because of his Malawian origins, as some openly said, *"Tingadzidza chinyi kumubwidi"* (What can we learn from this person of Malawian decent?). One of the villagers, Mr Fumayi says he saw no value in learning from Mr Phiri Maseko. Others openly confessed that Mr Phiri Maseko's innovations were labour-intensive although it ought to be noted that Mr Phiri Maseko's innovations might be labour-intensive as described in the previous chapters, they are constructed at the beginning and all that remains is maintenance work.

Envy also appears to be one of the reasons behind why some of the villagers have not adopted Mr Phiri Maseko's agricultural practices. During my fieldwork, I would hear comments about Mr Phiri Maseko from his neighbours such as "he has many friends from overseas" and "he receives money from his white friends". These comments are true but they are rather suggestive of envy. Even during the country's liberation struggle, for example, Mr Phiri Maseko was victimised after being allegedly sold-out by a neighbour for keeping an assortment of weapons left by freedom fighters at his home. The other reason could be that Hlupo and Ziyabangwa villages are geographically located near a *dekete* (wetland). The *dekete* is a busy place with most of the villagers doing market gardening, and collecting drinking water. These smallholders mainly produce vegetables on the *dekete* in contrast to smallholder farmers elsewhere in Zvishavane who have embraced Mr Phiri Maseko's practices.

In this study, I suggest that Mr Phiri Maseko was a frontiersperson, in Kopytoff terms (1987). Kopytoff asserts that:

> Contrary to a previously widespread stereotype of Sub-Saharan Africa as a continent mired in timeless immobility, its history has emerged to be one of ceaseless flux among populations...In brief; Africa has been a "frontier continent" – the stage for many population movements of many kinds and dimensions (Kopytoff, 1987: 7).

According to Kopytoff, a frontiersperson is the mobile African who transcends boundaries. As a person of Malawian origins, Mr Phiri Maseko managed to transcend frontiers and create spaces for other farmers. His innovations appealed beyond the confines of his horizons of Mapirimira Ward for he reached more than 1000 farmers in Zvishavane district alone (Wilson, 2013). Also, to further substantiate his status as a frontiersperson, his innovations reach an audience beyond Zvishavane as evidenced by farmers from many parts of the country who gave accounts (on the occasion of Mr Phiri Maseko's Lifetime Achievement Award) of how they have benefited from Mr Phiri Maseko's innovations (Wilson, 2010).

As a frontiersperson, Mr Phiri Maseko's innovations touched the hearts of many. There is an adopter's adopter in the Mazvihwa area of Zvishavane, a tomato farmer who said that he had never heard of Mr Phiri Maseko. Understandably, he might not have heard of Mr Phiri Maseko. Perhaps he does not have to know Mr Phiri Maseko to implement his water harvesting techniques. He would however, not have been harvesting water for market gardening if it were not for Mr Phiri Maseko. Ironically, such is the diffusion of his ideas that some adopters' adopters do not even know that their innovations can be traced to him.

There are also instances of smallholders who are using Mr Phiri Maseko's ideas to experiment and innovate further, these are his adapters. These include Mr Mugiya who lives in Mapirimira ward. His innovations include a humus-collecting basin at the foot of the mountain next to his fields. He collects the humus and adds it to cow dung in basins where he plants his maize, millet, and groundnuts. Mr Mugiya painted a picture of Mr Phiri Maseko (see Photograph 7). This shows the extent to which Mr Phiri Maseko became a mentor and an inspiration to his adopters and adapters. Mr Mugiya dug the 'Phiri pits' which have helped him to harvest enough for his family, and at the point at which we met, he was of the view that it was the time to take the 'might' of the established farmers. More of such innovative smallholders as Mr Mugiya are in Mazvihwa.

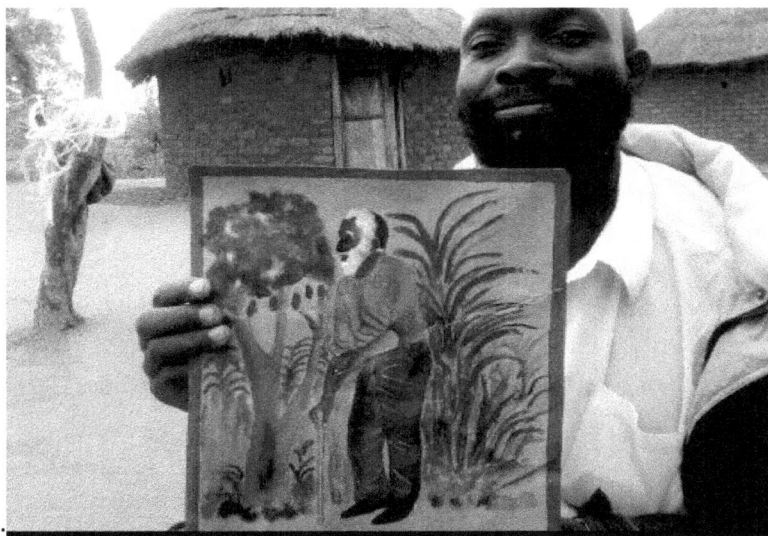

Photograph 7: Mr Mugiya, one of the adopters, with a painting of Mr Phiri Maseko

Mazvihwa is home to many of Mr Phiri Maseko's adopters and adapters engaged in different water harvesting techniques. They engage in path run-off harvesting. These smallholders harvest water that flows along paths in the villages. The water flows along small canals into small dams they constructed on their plots. Some of the smallholders have been assisted to design their dam walls by researchers such as Ken Wilson. Many in the area said to me that several youths have abandoned the 'gold rush'– illegal gold mining to do market gardening using Mr Phiri Maseko's ideas and methods as a form of risk aversion. Many youths are engaged in *chikorokoza* (gold panning) in the Zvishavane and neighbouring districts of Mberengwa and Shurugwi and some even make enough money to buy cars. The *makorokoza* (gold panners) are known for their lavish lifestyles during the few times they have windfalls. *Chikorokoza* is a hazardous preoccupation because it is characterised by the law of the jungle. Many youths have lost their lives because of the many squabbles that take place in the underworld where they 'mine' the gold. Some youths have realised that *chikorokoza* is not sustainable in the end. These youths are "commercialising livelihoods" by doing market gardening using water they harvest (see also Thomas et al., 2007). Mr Dube (see Photograph 8) did not hide his happiness about his Damascene moment - his 'conversion' from a *korokoza* (gold panner) to a water harvester:

> I am now making a living on farming because of Mr Phiri Maseko's ideas. *VaMamvura* gave me a new lease of life. Last year I used water from the small dam on my plot to grow tomatoes and I made about US $ 1500. *Chikorokoza* is now a thing of the past. I will never look back because I am now getting enough money to send my children to school. Here in Mazvihwa we are like lions roaring so as to guard our families from being 'attacked' by hunger (Interview with Mr Dube, 13 January 2013).

Mr Murefu who also tried his luck as a *korokoza* is also harvesting water that flows along village paths and has constructed a dam at his home (see Photograph 9). At the time of my fieldwork, he had grown tomatoes on his plot.

Photograph 8: Mr Dube (left) (a water harvester) and his friend stand in front of his small dam

Photograph 9: A small dam on Mr Murefu's plot with water used for watering his tomato crop

In Mazvihwa I also came across some innovative smallholder farmers (see Photograph 10) who have transformed gullies into resources. These smallholder farmers from Mazvihwa do not see gullies as a threat to the environment but rather as a resource. The villagers who are converting gullies into small dams always impress on visitors that *"Kuno kwaMazvihwa hativurayi nyoka, toparadza makoronga saka tinodzivirira kukukugwa kwevhu"* (Here in Mazvihwa instead of killing snakes, we kill gullies. We are erosion killers). Erosion killers engage in gully harnessing. They construct a dam wall along the gully and water moving through the gully is harnessed. Thus, the gully is transformed into a small dam and is used to water crops in their gardens. Ordinarily gullies are regarded as examples of land degradation in the rural areas. These smallholder farmers regard the gully as an opportunity to harvest water. This innovation by the erosion killers from Mazvihwa brings to the fore the efficacy of individuals' innovations. From my visits to farmers who say they duplicated Mr Phiri Maseko's strategies, I observed that some of them have gone a step further by doing their own innovations.

Photograph 10: Small dam (product of 'erosion-killing')

One of the most ardent adherents of the Phiri Maseko philosophy on the marriage of soil and water in the Mazvihwa area is Mr Banda of Mototi Ward; he is also of Malawian decent. Mr Banda mentioned that the most important lesson he learned from Mr Phiri Maseko, is that one must control water from the source. This is typical of what Mr Phiri Maseko did on his plot by constructing sand traps at the foot of the rock outcrop. Also like the water harvester himself, Mr Banda will never tire of saying that "when it is raining go outside and study the flow of water". Mr Banda has, through years of observation, managed to study his soil profiles and carefully chose sites for two small dams that were constructed in 1994. When I first visited him, the dams were full of water on his plot. He does market gardening as a way of adapting to climate variability. He has also bought livestock using proceeds from market gardening. During my last visit to his plot, he promised to give Mr Phiri Maseko a present of a goat as a way of appreciating the sustainable agricultural practices he leant from his mentor. While some farmers have adopted Mr Phiri Maseko's innovations in Zvishavane, it is important to acknowledge that there are hurdles to climate adaptation by smallholder farmers and I consider that in the following section.   .

Barriers and enablers to climate adaptation

Against seemingly insurmountable hurdles, Mr Phiri Maseko made inroads in building resilience to climate variability and thus inspiring other smallholders who live on the margins of survival. In this respect, he provided an enabling environment for uptake of innovations by other smallholders. The moral of his story is that navigating barriers to climate requires that we illuminate our blind spots (O'Brien, 2013). Illuminating our blind spots refers to confronting things we are still unwilling or unable to see (O'Brien, 2013). However, there is another equally important moral that flows from this story: merely learning from a successful innovator is not enough. Smallholders effectively adapt to climate by navigating around the labyrinth of barriers and utilising enablers peculiar to their own environment. Fuller (2008) echoes: "I am enthusiastic over humanity's extraordinary and sometimes very timely ingenuity. If you are in a shipwreck and all the boats are gone, a piano top buoyant enough to keep you afloat that comes along makes a fortuitous life preserver..." Mr Phiri Maseko's techniques have helped other smallholder farmers address climate variability in Zvishavane just like the 'piano tops' that Fuller contends are "fortuitous life preservers". Mr Phiri Maseko's agricultural practices can be viewed as offering an enabling environment for uptake of innovations.

As fortunes of smallholder farmers wax and wane in a changing climatic environment, it is important to reflect on what impedes or facilitates uptake of innovations. This study has alluded to traditional songs that play a pivotal role in informing Shona worldviews. These traditional songs create a fertile environment for uptake of innovations. During my fieldwork, I attended many field days that turned out to be platforms for smallholder farmers to display their innovations and share information through song and drama (see Photograph 11).

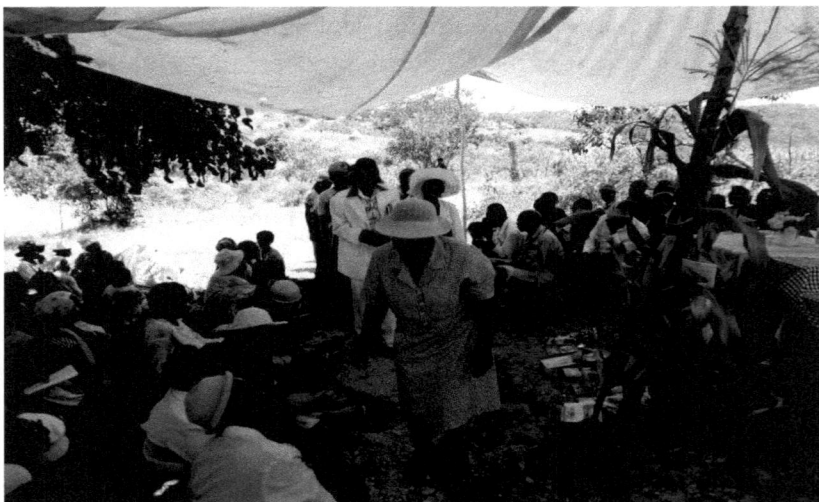

Photograph 11: Smallholder farmers performing a play sponsored by an NGO to drum-up support for the 'dhiga udye' programme

NGOs offer resources in the form of seed and inorganic fertiliser to the so-called lead farmers – Rogers' (2003) "early adopters" who are willing to experiment with CA. CA has met with mixed feelings in semi-arid southern Zimbabwe because the provision of resources appears to be important in smallholder farmers' decisions to embrace innovations. Moreover, giving inputs to 'lead' farmers (in most cases the village elites) leads to the further marginalisation of resource-poor farmers. The marginalisation of smallholder farmers can also be exacerbated by tensions in the villages, as evidenced by the power struggle in Ziyabangwa Village.

The power struggle between the two rivals for the village headship in Ziyabangwa has caused mistrust and this hinders information dissemination in the village. At the time when I left the field (at the beginning of 2013), Mr Phiri Maseko was trying to broker a deal between the two rivals who lead factions vying for control of the village, and which might have undermined the work of several NGOs in the community. What motivates NGO interventions in the rural areas is subject to debate.

Some NGOs target influential people such as councillors to help spread their message so as to influence uptake of innovations. The former Ward 6 councillor (see Photograph 12) helped influence villagers to implement NGOs' water harvesting techniques. The *kohwa mvura kohwa pakuru* t-shirts took Mapirimira ward by storm during my fieldwork. The water-harvesting message has also spread through farmers' clubs.

Photograph 12: Former Mapirimira Ward (6) councillor wearing his 'kohwa mvura kohwa pakuru' t-shirt

One of my respondents, Mr H Mutatiwa alluded to farmers' clubs that were formed in Zvishavane in the 1980s. Mr Phiri Maseko was the chairperson of one of the clubs named *Zarura Nzira* (open the way). These clubs motivated farmers to aspire to be *hurudza*.

However, some of the clubs are no longer functional. Reviving these clubs would help in up scaling of innovative farming practices that build resilience to climate, as demonstrated by Mr Phiri Maseko. Communities exposed to shocks - both natural and anthropogenic - have devised their own institutional responses to cope with, recover from and prevent future impacts (Bernier and Meizen-Dick, 2014). Institutional settings play a pivotal role in moulding the processes that are vital to innovation: interaction, learning and knowledge sharing (Hall, 2005). However, in instances where clubs are no longer active, individuals - such as ward councillors - fill the gap by holding meetings with other smallholder farmers in a bid to improve livelihoods.

Councillors oversee the distribution of government inputs such as seed and fertiliser but several have been accused of hijacking NGO sourced drought relief food for their own ends. In most cases, there are allegations that inputs do not reach the intended beneficiaries. Moreover, both government and ruling party officials mainly distribute maize seed. Such 'benevolent' gestures potentially exacerbate vulnerability as they promote crops (mostly maize) that are sensitive to climate variability at the expense of 'orphan crops' that are less sensitive (see also Smit, 1994). With Zimbabwe's multiple stressor environment worsened by a free-falling economy, it becomes tactical for villagers to 'align' themselves with the 'right' political party. Zimbabwe took a knock especially between 2000 and 2008 when the country's economy went into cardiac arrest, epitomised by a hyperinflationary environment unprecedented in modern history (see Table 2 below).

The decline of the country's economy also adversely affected extension services by AGRITEX in the rural areas. Most of the AGRITEX vehicles broke down and could not be repaired because spare parts were not available due to a serious foreign currency shortage in Zimbabwe in the 2000s. Resultantly, AGRITEX officials ceased visiting smallholder farmers (Interview with an AGITEX officer, Zvishavane 2012).

Year	Amount of ZWD equivalent to 1 USD
2001	55.99
2002	57.21
2003	575.49
2004	4 489.57
2005	21 531.19
2006	144 912.52
2007	9 467 500.35

Table 2: Decline of the ZWD between 2000 and 2007 (Source: Mutimukuru-Maravanyika, 2010: 48)

Rural communities faced other hurdles in the form of the breakdown of law and order especially in the volatile period from 2002 to 2008 when the ruling party ZANU PF's stranglehold on power came under intense threat from the opposition Movement for Democratic Change (MDC). Mutimukuru-Maravanyika weighs in:

Because of a breakdown of law and order in the country, many NGOs and international organisations found it increasingly difficult to work in politically volatile localities, especially the rural areas. The situation in rural areas was made worse by the passing of restrictive legislation such as the Public Order and Security Act (POSA) in 2002. Under POSA, anyone organising a public meeting is supposed to report to the police four days before the meeting to obtain a clearance (Mutimukuru-Maravanyika 2010: 47).

Such statutes as POSA greatly curtail the activities of food relief agencies such as the World Food Programme (WFP) (which distributes food relief to NGOs such as NGO 5) as they must seek clearance from the government. Approval often takes time because of the bureaucratic process involved. This means smallholder farmers take time to access the much-needed farming inputs such as fertiliser and seed provided by some of the NGOs. In other instances, NGOs withdraw prematurely before projects have taken off the ground, because of donor differences with the Zimbabwean government. This includes the ZWP, which has halted most of its operations due to the withdrawal of its donors. Withdrawal of the NGOs leaves rural communities even more vulnerable to stressors especially in the context of the country's economic decline. Due to the dollarisation of Zimbabwe's economy from 2008, many households' adaptive capacity has been compromised. The country now uses the US Dollar, South African Rand and the Botswana Pula as its currencies. The smallholder farmers who heavily depend on agriculture as a source of income can hardly access these currencies and find it difficult to buy agricultural inputs. Economic decline is one of the many stressors in a multiple stressor environment that has impacted negatively on the uptake of innovations by smallholders.

Uptake of innovations can be greatly compromised by failure to take into consideration the adverse effects of a multiple stressor environment especially the prevalence of the HIV/AIDS pandemic. Nguthi (2007: 201) notes that the major challenge is to "develop technologies that meet the evolving challenges of HIV/AIDS affected households without compromising productivity and sustainability of their livelihoods".

Uptake of Mr Phiri Maseko's innovations has been high where some NGOs might have failed because he 'rhymes' with other farmers as demonstrated by the farmers who have embraced his agricultural practices (Wilson, 2010). He situated water and soil in the same moral realm as other farmers do. In the Shona cosmology, water is life and Mr Phiri Maseko demonstrated this by innovations such as the sand traps which harvest and spread water in his water plantation. He had tangible results to show, for example, during the 2012/2013 rainy season when most smallholders' mango trees produced small amounts of fruit, the opposite was true for Mr Phiri Maseko. His trees bore heavily and people came from distances as far as Zvishavane town (about 15 km) to buy his mangoes. Mr Phiri Maseko's enviro-preneurship provided an enabling environment for adoption of innovations. His adopters in Mazvihwa such as Mr Banda and Mr Mawara planted fruit trees at their homes, and they also sell their produce and they in turn provide an enabling environment to their adopters (the adopters' adopters) in their area. Most of Mr Mawara's adopters are members of his church that is yet another enabling environment for uptake of innovations and information dissemination.

I held discussions with smallholder farmers who belong to the Methodist Church in Zimbabwe (see Photograph 13) and they suggested they find it easier to share information about farming with fellow Methodists because of the 'strong bond of trust' between them. From my fieldwork experience, I realised that some NGOs target church groups to market themselves. Church members find it easy to share information on agricultural practices that they see as effective in improving their livelihoods. During their meetings with smallholder farmers, some NGOs invite church leaders to preside over their functions.

Photograph 13: Smallholders, members of the Methodist Church stand next to the maize crop they harvested courtesy of CA

After considering the concepts of barriers and enablers to uptake of innovations, the section below discusses the various salient issues raised in this chapter.

Key issues at stake in the discourse of adaptation to climate

My findings demonstrate that smallholder farmers have many motivations to adapt including; exposure to drought, market shocks and increasing price of inputs. Other smallholders have much more exposure to climate variability compared to Mr Phiri Maseko because he implemented innovations that have enabled him to engage in several activities that have increased his adaptive capacity. He harvested water and this allowed him to engage in activities such as market gardening throughout the whole year because he had water available. Due to the availability of water, he grew reeds and sold them to basket weavers.

His 'water plantation' gave him the latitude to grow a variety of crops. For example, besides growing crops such as maize and groundnuts, Mr Phiri Maseko also grew sugar cane and a grove of bananas that have a readily available market in the form of neighbouring schools Msipane and Utongani. These activities show that he had less sensitivity to adverse effects of a multi-stressor environment than other farmers did. The farmers who have implemented his agricultural activities, for example those who 'rhymed' with him, the ones who constructed small dams on their plots, have also reduced their sensitivity to droughts, and increasing price of inputs.

That Mr Phiri Maseko 'rhymed' with other farmers is transformative. One of the characteristics of transformative initiatives is to be catalytic and upscaling and Mr Phiri Maseko's initiatives demonstrate this concept. His interventions are designed to meet the needs of other farmers including their vulnerability to a multi-stressor environment. Mr Phiri Maseko's agricultural practices are characterised by a sustainable agricultural low-external input system that resonates with other smallholders and achieves tangible results hence justifying the hard work and time invested. The moral from this observation is that unless architects of interventions are mindful of the needs of smallholders they will be unable to develop dynamic ways of assisting vulnerable communities. This perhaps explains why Mr Phiri Maseko's 'rhyming' with an audience succeeded where some outside interventions might have failed. Of significance, therefore, is that adaptation to climate change is not just about technical fixes but also addressing issues to do with drivers of vulnerability (Veteto and Crane, 2014).

As Mr Phiri Maseko 'rhymed' with an audience, it would be important to note that there are nuances to unpack. For instance, the ability to build networks is difficult to transfer since this is highly dependent on the ability of the farmer. In other words, Mr Phiri Maseko could not transfer skills of how he built his networks that have enabled him to access knowledge and finances he used in building his adaptive capacity.

Some may argue that Mr Phiri Maseko's transformative practices can exacerbate the vulnerability of other farmers downstream by preventing other smallholders from accessing water. However, because he understood the environment in which he was operating, he took measures such as constructing spillways that help to drain water downstream. One other positive from his innovative interventions is that the structures he constructed such as deepened contours help to minimise gully erosion. Minimising gully erosion is for the benefit of the community because in turn this minimises siltation of rivers. Moreover, he sunk wells on his plot and shared the water with his neighbours. Some villagers travel as far as 10km to fetch water on his plot.

His transformative adaptation can be unpacked by linking the concepts of conviviality and rhyming with his audience. That he is convivial means with other knowledge systems means that his ideas have resonated with other smallholder farmers who operate in the same realm as him. Mr Phiri Maseko's conviviality has enabled him to be innovative and thus building resilience to the negative effects of climate variability. Building resilience also means uptake of innovations from non-state actors.

Some of my findings agree to an extent with Rogers (2003). I found that some of the smallholders who implemented CA had a close relationship with extension officers, the so-called lead farmers. Other smallholder farmers only joined CA after witnessing proof of benefits. These are mostly smallholders who have other sources of income. However, it appears that uptake of CA is the lure of agricultural inputs especially when considered from the perspective of Zimbabwe's underperforming economy which has left smallholders more vulnerable to the adverse effects of climate variability and other stressors. On the other hand, Rogers' (2003) early adopters resonate with Mr Phiri Maseko's adopters, most of who adopted his innovations because of the promise of quick returns realised by harvesting water. Interventions that have a promise of early benefits - even though labour intensive - have a high uptake rate. Moreover, the question of quick returns is important given that the impacts of climate are changing quickly, and how the lure of benefits agrees with the Rogers (2003) model.

Rogers' model (2003) for adoption raises pertinent arguments in its explanation of the uptake of innovations. He argues that potential adopters consider potential for benefits. This agrees with my findings that some smallholders were lured to CA in the hope of accessing inputs. They were lured by the promise of inputs. On realising that inputs were only given to lead farmers; some swiftly withdrew from the project. Most of the smallholders adopt only if they can access inputs. In other words, most of the smallholders would be more interested in the inputs than the innovation. This suggests that CA creates dependence because it is input-intensive. On the other hand, water harvesting is a system that makes use of local resources, which although initially labour-intensive has two important results: (i) it builds capital; (ii) it stimulates creativity.

From my ethnographic study, I observed that CA entails that the farmer must harvest grass for mulching (see Photograph 14) almost every year because the possibility of the mulch (grass) being destroyed by termites is high. Most of the smallholders' homes are not fenced and cattle eat the mulch during winter when the cattle roam freely without being looked after by herd men. It is imperative to consider the likelihood of a conflict of interests between CA farmers and smallholders who keep livestock. Farmers who keep livestock (and do not engage in CA) are not happy that they are losing pastures for their animals to farmers engaged in CA (when they harvest grass).

Lack of inputs appears to be a major hindrance to adaptation to climate variability among smallholder farmers. Poverty pervades rural Zimbabwe thereby exacerbating the vulnerability of smallholder farmers.

There is need for a causal analysis of smallholders' vulnerability when considering interventions that assist smallholders to build resilience. Ribot adds, "making vulnerability analysis a required first step for any adaptation analysis or intervention can help move us from affirmative toward transformative climate action" (Ribot, 2011: 1161).

Photograph 14: Grass is used by CA farmers for the purposes of mulching

Moreover, the concept of structural vulnerability might help in adoption of innovations. I suggest that those that wield power such as the village elites, the *bambazonkes* in the rural areas, create barriers that hinder smallholder farmers from adapting to climate variability. For example, grabbing inputs meant for the marginalised supercharges a vicious cycle: Acts of greed by the *bambazonkes* inevitably lead to food insecurity among the smallholders. This increases the smallholders' vulnerability. Transformative change in the system could help more farmers access adaptive strategies. It is pertinent to find ways to support vulnerable groups. Scholars such as Tschakert et al., argue:

> A shifting discourse from vulnerability assessments to conditions for transformative change now calls for firm attention to the inequalities that undermine adaptive capacities, and methodological advances that expose social differentiation…We advocate for analyses that explicitly address structural drivers of vulnerability and their relational construction arising from inequality, marginalisation, poverty and constraining social–ecological dynamics while opening doors for transformative change (Tschakert et al., 2014: 341).

Inequality appears to be widespread in Zimbabwe's rural areas. However, vulnerability varies from household to household. Some households have access to remittances that help cushion them against shocks. Most smallholder farmers however, face a multi-stressor environment in the form of underlying institutional, political and economic barriers that hinder adaptive capacity. It is vital that climate change policies should consider the heterogeneous nature of smallholders' livelihoods. Heterogeneity includes consideration of belief systems among the smallholders.

Belief systems might help uptake of innovations or hinder up scaling. For insistence, some also accused Mr Phiri Maseko of using goblins to work for him. These smallholders have not replicated his farming practices perhaps for those reasons. From the aforementioned, I agree with Rancoli et al., (2009) and Ziervogel (2002) that belief systems play an important role in how individuals respond to climate variability. Taking cognisance of the belief systems can help make grassroots innovation more meaningful to smallholders.

The concept of innovation in agriculture refers to grassroots innovation. Letty and Bell (2012) suggest that grassroots innovation, in which smallholders participate, can be a complement to conventional research and development approaches. The data confirms with Letty and Bell's observations (2012) that grassroots innovation has higher rates of uptake as compared to linear models of technology transfer mainly because they are more applicable to the needs of smallholders. For example, water harvesting is transformative in the sense that it addresses the vulnerability of smallholders. Water harvesting, for instance, enabled Mr Phiri Maseko to engage in market gardening thereby increasing his adaptive capacity. This can be demonstrated by the smallholders in Zvishavane who have replicated his innovations. These smallholders are harvesting water and are engaged in market gardening, thereby boosting their livelihoods. A combination of water harvesting and organic farming has potential to be transformational, as Auerbach would say:

> Combining water harvesting with organic farming methods gives resource poor farmers strategy to improve soil fertility through raising soil colloidal humus levels, while sequestering carbon, increasing agro-biodiversity and raising nutrient and water holding capacity and also providing the means for entering high-end markets, domestically and internationally (Auerbach cited in Auerbach, 2013: 31).

As fieldwork wrapped-up, I mulled over my experience in Zvishavane. I began to re-think narratives of adaptation to climate. Is it not about time to take up the gauntlet of rural vulnerabilities by embracing smallholders' innovative ways of managing climate variability as a way of complementing external interventions in rural developmental issues? Is this not time for the "seventh sense" as Ramo (2016) says. In a seminal publication entitled, *"The seventh sense"*, Ramo examines "the historic force now shaking the world". Ramo deploys the "seventh sense" as a way of understanding the world. This study contends that the era of complexity we face needs a new approach, what Mabeza et al., (forthcoming) calls "business unusual". This study therefore sees a business unusual approach as the "seventh sense". Water harvesting tailored on smallholder technologies offers a pathway of adapting to climate and is in a way a business unusual approach.

This study recognises that local innovations have been incorporated in rural development projects before, but points out that in some cases the development agents do not want to let go of the steering wheel so that the smallholders are actively involved in issues to do with their destiny. As already mentioned, some development agents are at times in the 'backseat' barking orders to smallholder 'drivers'. This book demonstrates that in practice there has been a 'marriage' of local practices and interventions by smallholders and development agents. Waters-Bayer and van Veldhuizen concur, "In rural development the major challenge is to move beyond the existing innovations farmers have developed using IK and creativity, and to develop these ideas further in joint experimentation, integrating relevant information and ideas from elsewhere" (Waters-Bayer and van Veldhuizen, 2004: 1).

As an innovative smallholder farmer, Mr Phiri Maseko's agricultural practices have helped smallholder farmers in semi-arid environments to adapt to a changing climate.

Conclusion

Whether as a *mubwidi* or an award-winning smallholder, Mr Phiri Maseko left a legacy in the history of smallholder innovations in semi-arid environments. The discussion on Mr Phiri Maseko is not just about being a *mubwidi* and award winning *hurudza* but his broader meaning in an era characterised by increased climate variability and uncertainty. The farming practices initiated by Mr Phiri Maseko in rural Zvishavane offer potential for building adaptive capacity to rainfall variability in semi-arid regions. The material in this chapter suggests that local innovations such as those of Mr Phiri Maseko have potential for sustainability. Agriculture is sustainable because of hard work by smallholders. High input systems such as CA have challenges because smallholders do not have financial resources to purchase inputs and at times, the inputs do not get to them on time.

This study cautions that it would be incorrect to suggest that the Phiri Maseko adopters' success is solely attributed to their mentor (Mr Phiri Maseko), because they have demonstrated innovativeness among themselves. Many smallholder farmers who 'rhyme' with Mr Phiri Maseko have benefited from implementing his water harvesting techniques as evidenced by their grain yields.

"Why should we deepen our contours?" asked the boy whom I quoted at the start of this chapter. With the need to manage rainfall variability and boost food security in semi-arid southern Zimbabwe, the more critical question is why any smallholder farmer wouldn't?

Conclusion

Good news makers stand and be counted

> I learned to redefine myself regardless of what happened to me when I was a kid…I have been able to reclaim myself. This is something that is required for every individual. We are not what happened to us (Chido Govera cited in Smith, 2014).

Introduction

The travails of smallholders open a window in the climate change discourse usually overshadowed by hegemonic narratives. This is evident in the example of Mr Phiri Maseko who assumed the role of the astute spokesperson for the 'stifled' constituency of rural farmers in Zimbabwe. The Phiri Maseko story - that of water and soil in holy matrimony – is a *good* news story, a clarion call to challenge the shibboleths of doom mongers and players in the developmental discourse. Rural developmental thinking is mainly premised on the us-versus-them chauvinistic binary, where smallholders are mostly seen as passive recipients of technologies, typical of the pipeline model. In contrast, this study argues that the marginalised's elegy gives way to hope as evidenced by the "*good* news" story of Mr Phiri Maseko's remarkable innovative practices that this study presents.

Thus, this research has not been intended to be a quantitative study, but a qualitative ethnographic study that seeks to understand an issue differently in order to phrase questions differently, it seeks to understand what counts in the rural development discourse, rather than seeking to count (to cite Nyamnjoh, 2013b). The study has much wider relevance to southern Africa, and is grounded in regional philosophy and practice of water and soil. It is a product of a study of Mr Phiri Maseko, a transcendent smallholder farmer who harvested water, a way of redefining adaption to climate variability. This is what this study terms the "seventh sense". Ethnography offers a front row seat to lived realities of smallholder farmers in a changing climatic environment. Fourteen months of fieldwork in rural Zimbabwe afforded me insights into Mr Phiri Maseko's innovations for managing climate variability. The conclusion discusses the salient points I raise in this study that draws on the concepts of vulnerability, resilience, adaptation and innovation to probe smallholder pathways for managing climate variability. This study demonstrates that Mr Phiri Maseko possessed a high adaptive capacity. His agricultural practices promote the concepts of conviviality and creativity that helped him to boost his adaptive capacity. This book suggests that local innovations such as those of Mr Phiri Maseko can play a role in adaptation to climate variability, more so in an era of uncertainty that is difficult to predict. Thus, this study suggests that what is predictable about climate change (a complex issue) is its unpredictability. Such unpredictability warrants a business unusual approach.

Mr Phiri Maseko's disintermediation is business unusual. The pro-truthers' innovations (premised on the rejection of dominant narratives) could help to try to address this complex issue. However, innovations by smallholder farmers neither are a silver bullet solution nor are they a gigantic leap in innovative progress. Rather, they offer a complementary role (premised on appreciation of local dynamics) to exogenous efforts that try to address perennial food insecurity among smallholder farmers in the face of climate variability. It is out of this realisation that in the Conclusion, I assert that Mr Phiri Maseko built resilience to climate variability by becoming a latter-day *hurudza*. I suggest that the approaches based on dominant narratives aimed at boosting food security in the rural areas ought not be rejected but used in tandem with smallholder farmer innovations in the discourse of rural development. However, water harvesting alone may not suffice. Mr Phiri Maseko appeared to realise this as evidenced by how he converted his car into a taxi that operates in Zvishavane town (taking advantage of the spendthrift gold panners) so as to diversify his livelihood options. Smallholders should be encouraged to try to diversify their livelihood options in order to adapt to their variable environments.

Reflections

The study argues that Mr Phiri Maseko built resilience by redefining climate adaptation in semi-arid rural Zvishavane. Smith's (2014) article on how a mushroom grower, Chido Govera redefines survival strategies in the face of adversity is instructive. Govera orphaned at the age of seven (Smith, 2014) conjures ingenious ways of reducing vulnerability by growing mushrooms so that she adapts to a multi-stressor environment. Likewise, Mr Phiri Maseko has redefined adaptation to climate variability by his own unique way of innovating that marries endogenous and exogenous practices, thus challenging conventional barriers. Skrydstrup (2009: 357) posits, "resilience is not necessarily endogenous; resilient societies are not alone and never have been". This study therefore does not aim to discard scientific knowledge and replace it with local sources of knowledge but advocates for a complementary interplay between different knowledge systems. Mr Phiri Maseko, for example, used both hybrid seed bought from seed houses and his own varieties, which are open-pollinated. Local knowledge is equally important as learning from others. Yet this remains an ongoing challenge for NGOs (who mainly support external knowledge) to roll out at an appropriate scale locally. Weaving together different knowledge systems, as shown by Mr Phiri Maseko, is advisable, but heavily time-dependent. Water harvesting is at the heart of his agricultural practices. The value of water harvesting is such that initially the process might be labour intensive, however in the end it reduces the need for capital, stock or infrastructural inputs. The water harvesting structures are built during the first phase of implementation and thereafter the smallholder carries out maintenance work. In contrast, NGO-led interventions mainly require a continuous supply of inputs.

Harvesting or 'domesticating' rainwater, is a system that reduces the incidence of soil being washed away. Domesticating rainwater resonates with Mr Phiri Maseko (of Malawian descent) who ironically was also 'domesticated' by Shona cosmologies. By being 'domesticated' by the Shona, Mr Phiri Maseko's farming practices reflect Shona conceptualisation of water, soil and marriage, among other influences. Interdependence of different parts is at the heart of his conceptualisation of water and soil. This study argues that the ideas of interdependence and conviviality are the cornerstones of building resilience to climate variability. Interventions based on this realisation are likely to yield positive results ensuring a high uptake among the smallholder farmers. What is clear from the ethnographic data presented here is that farming practices are informed by a realisation that dualisms are problematic.

Mr Phiri Maseko's water harvesting practices can play a pivotal role in ensuring food security and poverty alleviation. Mr Phiri Maseko, through his marriage of water and soil made an important contribution to rural livelihoods in southern Zimbabwe. In this study, I argue that he boosted his food security and regenerated his local environment through his farming techniques in an area characterised by erratic rainfall. His innovations make the scientific community re-think the notion of what successful adaptation to changing climate may entail.

I argue that Mr Phiri Maseko realised that one's environment should spur one to be innovative. I give an example of how he experimented by planting a mango seedling despite the belief then that the moment the mango tree bore fruits the one who planted it would die. Such experimentation made him a *hurudza*. Smallholder farmers are relentless experimenters in the face of a multiple-stressor environment, and thus the innovative process is continuous. Such innovations find easy acceptance by other smallholders because the technologies are simple and inexpensive, and thus easy to replicate. These technologies straddle barriers of indigenous and exogenous knowledge systems. These technologies help smallholders build resilience to increased rainfall variability.

I argue that smallholder interventions need not only be informed by lessons from experiences from the past to plan for the future, they should also incorporate future climate projections. However, the rapid rate at which climate impacts are changing means smallholders are not able to rely on the past alone but must move towards transformational adaptation that address vulnerability of those in the frontline of the adverse effects of climate variability.

An enabling environment is vital to smallholders' adaptation to changing climate. Mr Phiri Maseko, for example, utilised the many sources of knowledge such as his physical environment, his friendship with researchers Ken Wilson and Ian Scoones in addition to institutions of learning such as Makoholi Agricultural Institute and the informal learning environment at the 'University of Mototi'. These sources of knowledge provided an enabling space for learning. Ziervogel and Calder are emphatic about the need to utilise enabling conditions, "If adaptation to climate variability is a priority, it is the enabling conditions that need to be explored..." (Ziervogel and Calder, 2003: 415).

The *hurudza* concept of the pre-colonial era as epitomised by Mr Phiri Maseko can help provide an enabling environment. Smallholders are likely to improve their livelihoods by implementing tried and proven agricultural practices such as those of *hurudza*. *Uhurudza* entails the need to negotiate barriers. Latter-day *hurudza*, such as Mr Phiri Maseko, presided over many 'marriages' that accommodate different knowledge systems. Mr Phiri Maseko was an "enviro-preneur". His successful growth of mangoes was made possible by his water harvesting practices. Water harvesting practices, earned him income every year. Photographs in this study demonstrate his agricultural practices and crops he grew for sale are testimony to his enviro-preneurial skills. This study advocates for smallholders to embrace the latter-day *uhurudza* concept in the mould of Mr Phiri Maseko, which entails enviro-preneurship. Latter-day *uhurudza* is transformational at both the individual and community level. *Uhurudza* enables the farmer to purchase assets to reduce vulnerability.

Mushongah (2009) suggests that vulnerability should not only be seen as the identification of threats, but also as the means of resistance. Moser (cited in Mushongah, 2009: 305) adds that, "vulnerability is therefore closely linked to asset ownership. The more asserts people have, the less vulnerable they are, and the greater the erosion of people's assets, the greater their insecurity". This study documented Mr Phiri Maseko's assets that he bought using proceeds from his *uhurudza*. One such asset is the car he bought and converted into a taxi whose proceeds have helped him adapt to a multi-stressor environment.

At the heart of enviro-preneurship therefore, is building adaptive capacity that in turn reduces vulnerability. This is critical given that the research findings show that lack of financial resources exacerbates smallholders' vulnerability to rainfall variability. Smallholders should move from sole subsistence farming to income generation as enviro-preneurs. An enviro-preneur such as Mr Phiri Maseko was innovative and this helped him boost his adaptive capacity to impoverished conditions. For example, the Phiri Maseko family generates money by selling fruits such as mangoes farmed on its plot.

Mr Phiri Maseko and the *chinamwe chiri mumusoro* smallholder built resilience to climate variability. Their innovations are part of a wider picture of how rural communities innovate to reduce vulnerability. For instance, I recall my childhood in the village when smallholders would organise *humwe* (work parties) to ferry anthill soil to their fields as a way of adding fertility to their pieces of land. This and many other ingenious ways helped them to boost their food security by producing enough grain to feed their families and thus catapulted them to *uhurudza* status. If smallholder farmers show such determination as demonstrated by Mr Phiri Maseko, they may be able to overcome barriers that impede their adaptation to climate. Thus, innovations that help build resilience to climate are partly informed by an appreciation of local dynamics in one's own environment.

For most of the smallholder farmers who live in variable environments, innovations such as those of Mr Phiri Maseko could be appropriate. His innovations have been embraced by some smallholders due to their proven success in boosting food security and access to water. The innovations are cost-effective and although they require labour, they nevertheless offer smallholders a route to self-reliance. Self-reliance is synonymous with Mr Phiri Maseko's latter-day *uhurudza*. As a latter-day *hurudza*, he was self-reliant since for food requirements he depended on crops he grew on his plot.

Mr Phiri Maseko did not only 'export' produce from his plot to other areas of Zvishavane but also exported his ideas which have potential to boost smallholders' productivity. Because his ideas spread, he managed to establish himself as an opinion leader in Zvishavane. Most of the people in his ward looked to him for advice and help, and his influence was palpable during fieldwork. During the research period, the rainy season was characterised by erratic rainfall, some smallholder farmers harvested low amounts of grain. Yet, despite the drought, Mr Phiri Maseko managed to harvest enough grain to feed his family. Many of his neighbours who had harvested less grain would come to ask food at Mr Phiri Maseko's home. One regular visitor who came for food at Mr Phiri Maseko's home was nicknamed the "boarder" by the Phiri Maseko children because he looked like he had taken up residence at the Phiri Maseko home. I came to realise that the 'boarders' at his home were many. Other boarders – non-humans have also 'moved-in' at the Phiri Maseko residence and these include a zebra, pigeons, and goats. Building resilience not only benefits the innovator and his/her family, it also benefits other components in the socio-ecological system such as livestock and even wild animals.

Mr Phiri Maseko managed to send his children to schools that offered better education because he could afford to pay the fees. He sent one of his daughters to a renowned agricultural college near Harare. In the 1980s when ownership of cars in Zimbabwe was still considered the preserve of a few, Mr Phiri Maseko bought his first car.

His innovations suit other smallholder farmers' needs. Mr Phiri Maseko's agricultural innovations rhyme with those of other smallholder farmers in rural Zvishavane. His audience has warmed-up to his agricultural practices with enthusiasm. This can also be attributed to the fact that smallholder farmers adopt adaptive strategies that show tangible results. His adopters have become keen water harvesters and some are involved in market gardening thereby greatly enhancing their statuses as enviro-preneurs. This study therefore asserts that innovations such as those of Mr Phiri Maseko that demonstrate tangible results and promote independence and innovation have high uptake among smallholder farmers.

Uptake of Mr Phiri Maseko's innovations can partly be attributed to what appears to be the government's abrogation of its duties of helping uplift smallholder farmers. The role of the state as a major actor in environmental governance is waning (Agrawal and Lemos, 2007; Scoones and Thompson, 2009). Agrawal and Lemos (2007) suggest that the "shrinking state" is characterised by tax cuts and privatisation, and this implies that governments allocate very few resources to environmental issues thereby compounding the vulnerability of smallholders to rainfall variability leaving the rural space open for interventions by NGOs.

NGOs and enterprising individuals such as Mr Phiri Maseko filled that gap left by the government. NGOs view this as an opportunity to help spread their influence in the countryside and thus have become very important players in the improvement of rural livelihoods. Some NGOs in Zvishavane have embraced Mr Phiri Maseko's water harvesting techniques and are using the *kohwa mvura kohwa pakuru* (harvest water for a bumper crop) slogan. Pertinent to note is that such NGO interventions that embrace local practices are embraced by smallholder farmers; thus, demonstrating lateral collaboration. NGOs ought to not only be viewed through negative lens. My study argues that NGOs in Zvishavane have assisted smallholders in building resilience through provision of inputs.

At times, however, projects by various players in the rural development discourse (politicians included) often fall prey to what Kay (2009: 11) has called 'adaptation by ribbon cutting'. 'Adaptation by ribbon cutting' results in the projects collapsing because mostly they are externally conceived and not premised on the needs of the recipients but on the goals of the funder. For example, NGO projects are strictly structured by distinct deadlines. Adaptation by ribbon cutting is a sly wink at the political establishment, who mainly view interventions in rural areas in terms of political capital.

Uptake of innovations by smallholders is mainly influenced by the expectation of benefits. However, some smallholders wait until they see tangible benefits. Subsidising agricultural inputs will potentially aid smallholder farmers boost food security. Failure to access inputs compromises smallholders' capacity to feed themselves.

Boosting food security was at the heart of Mr Phiri Maseko's *uhurudza*. However, he also used *uhurudza* to gain recognition in an area with predominantly Shona speaking people given that people of Malawian origins are looked down upon by some sections of the Shona societies. Nevertheless, he was a frontiersperson (Kopytoff, 1987) who appealed beyond his Malawian descent. He also appealed beyond his village and country. Mr Phiri Maseko straddled generations as evidenced by some of the smallholder farmers in their 20s who are replicating his innovations. *Uhurudza* catapulted him into a revered change-maker. I view his innovations as part of a suite of practices that can help smallholders adapt to climate variability.

I argue that accommodating both creative local agricultural practices and ideas from research institutions can help smallholder farmers to build resilience to climate variability. In this era of complexity in the wake of global climate change, it is important to embrace creative innovations by smallholder farmers, which have shown promise.

Effective innovation must co-evolve with a changing world (Pelling, 2011). Fenn (quoted in Wall, 2014) says: "But you cannot afford to stay still - business is a moving escalator. The world is moving around you - customer expectations are changing…" The world is characterised by mobility and hence innovations tend to be a fusion of ideas from different localities. Innovators such as Mr Phiri Maseko continuously innovated for their innovations to remain relevant. Innovations such as these withstand the test of time and are continuously upgraded to suit changing times. Thus, interventions exhibit an element of dynamism. Such interventions encompass many forms of knowledge systems.

Local innovations such as those of Mr Phiri Maseko have helped other smallholders to adapt to changing climate. In this study, I documented farmers who have replicated Mr Phiri Maseko's innovative adaptation to climate variability. These farmers comprise the audience that 'rhymed' with him. The ZWP is one such platform where farmers can share their experiences and elect to 'rhyme' with Mr Phiri Maseko. Rhyming with Mr Phiri Maseko spurred smallholder farmers to try to be innovative in their own settings. The smallholder farmers experiment in their environments so that they try to manage rainfall variability. This study gives an example of the 'erosion killers' of Mazvihwa whose innovations are aimed at addressing gully erosion by converting gullies into small dams – a process (with their characteristic flair for humour) where they say they "don't kill snakes" but "kill erosion".

Innovations premised on an appreciation of the local ecology have potential to assist smallholders when compared to dependence on expensive external inputs. The persistence of dominant epistemologies results in certain agricultural techniques being given precedence at the expense of 'others'. Such "blind spots" as Tiffen (1996) would say, are very pervasive and they fail to 'see' such tried and tested soil and water conservation techniques as those of Mr Phiri Maseko.

Such 'crisis narratives' of so-called environmental destruction are driven by science's 'snapshot' methods which result in a flawed diagnosis of environmental problems and this might lead to maladaptation. Maladaptation inadvertently leads to increased vulnerability. This study calls for self-introspection on the part of the smallholder farmers, government and non-state actors working to assist rural communities build resilience to climate variability.

The need for change of mindsets is evident (as stated elsewhere in this study) in Mr Phiri Maseko's village where some smallholder farmers suspect that he uses *divisi* and goblins to produce a lot of grain. Some of the villagers' beliefs have become a barrier for uptake of innovations. Rethinking these values and beliefs will assist the smallholder farmers to adapt to climate variability. Admittedly, changing mindsets is a difficult proposition. However, other farmers who have replicated Mr Phiri Maseko's system can also help in doing extension work, to help show other smallholder farmers the benefits of such farming practices as water harvesting. Smallholders can benefit from policies that recognise the important role their innovations play in rural development.

National policymakers appear not eager to empower the rural people to facilitate adaptation processes that take local knowledge into account. Policies should recognise the diversity of local innovations and create an enabling environment for effective local adoption. Policy makers are mainly influenced by science in making decisions that they think might reduce rural communities' vulnerability. These decisions however, create many problems that negatively affect rural people. As the saying goes, at times problems are caused by solutions. Interventions should have the capacity to address the concerns of smallholders. Policies that are largely informed by dominant exotic knowledge systems at the expense of local ones might not achieve the object of building smallholder farmer resilience.

In comparison, Mr Phiri Maseko's interventions were on going and subject to modification to ensure longevity. Policies premised on this realisation of smallholder needs helps smallholders adapt to a changing climatic environment. It puts the last (in this case the rural poor) first (Chambers, 1983). Such approaches that give prominence to the rural poor on the climate change agenda could tap and release the latent energy and talents of the rural communities. Rural people should also play a role in the innovation process as they have knowledge of their forms of production and social context if they are to manage complexity associated with the unpredictability of climate.

In light of this unpredictability, this study suggests a "business unusual" approach, an appeal to the "seventh sense", a new way of trying to understand the complex issue of climate change. No doubt, recent events in the United States of America, the advent of *Trumpism* (policies initiated by Donald Trump that advocate the rejection of the political establishment) have set tongues wagging. How has the establishment failed the aspirations of many? Should the "usual-usual" approach be relegated to the dusty drawers of history? This is a wake-up call for stakeholders in the climate change discourse that sets out a compelling vision of reassessing adaptation interventions. Clearly, the usual, "usual-usual" approach has not achieved desired results. Now is the time to do things differently if we are to manage complex problems that afflict humanity today.

Conclusion – Why are they still 'there'?

"Why are we still here?" This question posed by Andy Hall has reverberated throughout this study. This book departs from conventional wisdoms in developmentalist literature – the sort of wisdoms that are inadequate and that have led to 'farmer first' conferences still struggling with the same questions after two decades. The work shows an alternative: grounding agricultural infrastructural innovation, in the form of pits, canals, etc. in regional thought/philosophy (for example, Maathai's 'foresters without diplomas' (2006) and Bloch's (1996) story about Robert Mazibuko) and Nyamnjoh's (2002) concept of conviviality to include rationality across households, species, water soil and generations. The study also gave examples of farmers in East Africa whose conceptualisation of water and soil resonates with the regional agricultural practices. The study argues for a regional philosophy water and soil that is a profoundly rich resource for innovation and experimentation rather than a static approach to local knowledge that does not allow for sufficient appreciation of the ongoing experimentation and innovation of regional farmers. Such appreciation of a regional philosophy of water and soil is the "fierce urgency of now".

I address Andy Hall's question by building on the experiences of Mr Phiri Maseko and how his innovative farming practices that shook the bedrock of extension services, boosted his adaptive capacity in climate variable southern Zimbabwe. This study's major thrust has been to rethink adaptation to climate by focusing on the innovative farming practices of Mr Phiri Maseko. To that end, the study used lens of resilience and argued that the farming practices of Mr Phiri Maseko are premised on the realisation that dualisms are problematic. They are informed by the need to harvest water as a way of adapting to erratic rainfall in semi-arid rural Zimbabwe. Building resilience to climate as demonstrated by Mr Phiri Maseko is not wholly endogenous. Rural communities fuse endogenous practices with exogenous ones, a strategy that is at the heart of conviviality. In other words, adaptation "from the inside-out" (O'Brien, 2012) a self-introspection by the different stakeholders that calls on smallholders facing climate variability to 'marry' various forms of knowledge that embrace the reality of interdependence in the socio-ecological system, is imperative.

This study does not at all aim at earning kudos from latter day Luddites by appearing to be at odds with the role of technology in rural development. Neither is the study shouting from the rooftop: "not so fast" to the techno-ebullience and the "Buckminster Fullerish optimism of transhumanists". Instead, the study argues that the pipeline model fronted by its proponents who have "learned nothing and forgotten nothing" has failed. It is time for 'old dogs' to drool at the prospect of new tricks, for example, by embracing Mr Phiri Maseko's hybrid concept. Thus, his hybrid concept provides one-stop shopping for stakeholders in the rural development discourse who wish to address smallholder perennial food insecurity.

In addition, this study has argued that for uptake to occur, innovations need to take into consideration the needs and knowledge of smallholders. The study asked: "Why are they (smallholders) still there?" in reference to the vulnerability of smallholder farmers to a multi-stressor environment. If smallholder farmers in dryland areas harvest water as a way of building resilience to climate variability, they may not remain 'there', trapped in vicious cycles of food insecurity. Agricultural practices that demonstrate tangible benefits and are cost-effective such as the *uhurudza* of Mr Phiri Maseko (that hinge on water harvesting and soil fertility) have high uptake among smallholder farmers. As Mr Phiri Maseko said, he learned to rhyme with his environment and other smallholders. Smallholder farmers face difficult hurdles in a bid to build resilience to climate variability. However, amid these challenges, Mr Phiri Maseko's farming practices for managing rainfall variability (his marriage of water and soil in holy matrimony) redefine adaptation to climate and offer green shoots of hope to dryland smallholder farmers in Zvishavane and in similar environs. Such an example of a good news story is what we need today instead of *bad* news stories of "imminent peril". Therefore, this study calls on *good* news makers to stand be counted.

Appendix 1 - A tribute to the late Mr Phiri Maseko*

For whom the bell tolls

"Days are numbered", Mr Zephaniah Phiri Maseko would say, as his biological clock's ticking grew louder by the day. Sadly, and unknown to him, 1 September 2015 was to be the last of the numbered days, the day he took his last breath. The internationally acclaimed water harvester from Zvishavane, rural Zimbabwe passed on after succumbing to an illness at the advanced age of 88. I was privileged to occupy a front row seat to his agricultural practices during a fourteen-month long ethnographic study for my PhD at his residence a few years ago when I learned many things about him. He is the smallholder innovator who taught us to 'marry water and soil', 'plant water' as we plant crops, and conserve our environment. Mr Phiri Maseko's innovative agricultural practices have helped smallholder farmers in semi-arid regions to adapt to a changing climatic environment.

Born in 1927 in Zvishavane, he grew up in colonial Zimbabwe in a harsh environment characterised by rainfall variability. As if that was not enough, he had to contend with poor soils and oppressive legislation. In the 1970s, he was arrested by the colonial authorities for daring to challenge legislation that forbade him from cultivating a wetland. Ultimately, he was allowed to till the wetland after authorities realised that his agricultural practices were sustainable. His agricultural practices included construction of structures that harvest water.

With the advent of the Zimbabwe's liberation struggle in the 1970s, soldiers of the Rhodesian army placed Mr Phiri Maseko under house arrest after an arms cache was 'discovered' at his home. He was placed under house arrest until independence in 1980. In the post-independence era, he continued to upgrade structures that harvest water that include what he called sand traps and the 'immigration centre' where he 'welcomed' water to his plot. His adroit water harvesting techniques transformed his plot into a 'Garden of Eden'. These techniques enabled him to adapt to climate and to date his agricultural practices have spread to other areas of Zvishavane and beyond. He shared his practices with other farmers mainly through the non-governmental organisation, Zvishavane Water Project that he helped to find.

With Mr Phiri Maseko's death, John Donne's question: For whom does the bell toll, sounds fitting. Donne says that all of us are part of humankind and that any person's death is a loss to us all and therefore the bell tolls for all of us. The bell of Mr Phiri Maseko's death tolls for us all. We should pause and reflect on his legacy and how we have benefited from his agricultural practices as we grapple with an uncertain future. Greenhouse gases continue to be 'pumped' into the atmosphere. There appears to be no agreement in sight for the reduction of greenhouse gas emissions. The major emitters continue to bicker, typical of a proverbial dialogue of the deaf with nobody appearing to comprehend what the other is saying. In these trying times, we take solace from the opportunities that come our way if we embrace innovations for managing climate variability from smallholders such as Mr Phiri Maseko.

We should consider ourselves fortunate to have lived in his era. The real meaning of Mr Phiri Maseko is not what he has left his children with but what he has left in them and indeed in all of us – that in our individual efforts we can work to conserve the earth that we all depend on for survival. In the Chewa language of Mr Phiri Maseko's ancestry, all we can say is *"zikomo"* (thank you) (as Ken Wilson, his friend would say). Your bell tolls for us all.

*I first wrote this article for the Amanzi for Food project, Rhodes University, South Africa.

List of references

Achebe C (1974) *Arrow of God.* London: Heinemann.

Adger W N (2000) Social and ecological resilience: Are they related? *Progress in Human Geography* 24: 347 DOI: 10.1191/030913200701540465.

Adger W N (2006) Vulnerability. *Global Environmental Change* 16: 268-281.

Adger W N, Dessai S, Goulden M, Hulme M, Lorenzoni I, Nelson D R, Naess L O, Wolf J and Wreford A (2009) Are there social limits to adaptation to climate change? *Climate Change* 93 (3-4): 335-354.

Adichie C (2009) "The Danger of a Single Story," Talk filmed July 2009, at Technology Entertainment Design (TED) Global. Available at: http://www.ted.com/talks/chimamanda_adichie_the_danger_of_a_single_story.html.

Agrawal A and Lemos M C (2007) A greener revolution in the making? *Environmental Governance in the 21ˢᵗ Century* 49 (5): 36-45.

Aidenvironment Ecole Polytechnique Federale De (EPFL) (2013) Defining Smallholder, suggestions for a RSB smallholder definition. Lausanne, Switzerland.

Alliance for a Green Revolution in Africa (AGRA) (2014) *African agriculture status: Climate change and smallholder agriculture in sub-Saharan Africa.* Nairobi, Kenya.

Amutabi M N (2006) *The NGO factor in Africa: The case of arrested development in Kenya.* New York and London: Taylor and Francis Group, LLC.

Anwar D (2014) Can Alexis Sanchez have the "Luis Suarez-effect" at Arsenal this season? Available at: www.squawka.com/news.

Arnall A, Kothari U and Kelman I (2013) Politics of climate change: discourses of policy and practice in developing countries. *The Geographical Journal* 180 (2): 130-140.

Auerbach R (2013) Transforming African agriculture: Organics and Alliance for a Green Revolution in Africa (AGRA). In: Auerbach

R, Rundgren G and Scialabha N El-Hage (eds), *Organic agriculture: African experiences in resilience and sustainability.* Food and Agriculture Organisation of the United Nations (FAO), Rome, 16-34.

Babbie E and Mouton J (2001) *The practice of social research.* Southern Africa: Oxford University Press.

Bahadur A V, Peters K, Wilkinson E, Pichon F, Gray K and Tanner T (2015) The 3 As: Tracking resilience across braced. Soapbox, UK. Available at: www.odi.org/sites/odi/.org.uk/files/odi-assets/publications-opinion-files/9812.pdf.

Baldwin J (1965) *Sonny's blues. Going to meet the man.* New York: Dial Press.

BBC News (2016) 'Post-truth' declared word of the year by Oxford Dictionaries. 16 November. Accessed at: www.bbc.com/news/uk-37995600

Berggren N and Bjørnskov C (2009) Does religiosity promote or discourage social trust? Evidence from cross-country and cross-state comparisons. A paper presented at the social dimensions of religion in civil society at Ersta Sköndal University College in Stockholm.

Bernier Q and Meizen-Dick R (2014) *Networks for resilience, the role of social capital.* Washington D C: International Food Policy Research Institute.

Berrang-Ford L, Ford J D and Paterson J (2010) Are we adapting to climate change? *Global Environmental Change* 21 (2011): 25-33.

Bird-David N (1993) Tribal metaphorisation of human-nature relatedness. In: Milton K (ed), *Environmentalism: The view from anthropology.* London: Routledge, 111-124.

Birdsall N (2011) The global financial crisis. The beginning of the end of the 'development' agenda? In: Birdsall N and Fukuyama F (eds), *New ideas on development after the financial crisis.* Baltimore: The Johns Hopkins University Press, 1-23.

Blackmore C (2007) What kinds of knowledge, knowing and learning are required for addressing resource dilemmas? A theoretical review. *Environmental Science & Policy* 10: 512-525.

Bloch J (1996) *The tree man: Robert Mazibuko's story*. Pietermaritzburg: New Readers Publishers.

Blommaert I and Jie D (2010) *Ethnographic fieldwork, a beginner's guide*. Clevedon: Short Run Press.

Bologna S (2008) *Nature and power: A critique of 'people-based conservation' at South Africa's Madikwe Game Reserve*. PhD Thesis, University of Cape Town, South Africa.

Bourdillon M F C (1987) Guns and rain: Taking structuralist analysis too far? *Africa: Journal of the International African Institute* 57(2):263-274.

Boyd E. and Cornforth R J (2013) Building climate resilience in Africa: Lessons of early warning in Africa. In: Moser S C and Boykoff M T (eds), *Successful adaptation to climate change. Linking science and policy in a rapidly changing world*. London: Routledge, 201-219.

Brooks N (2003) Vulnerability, risk and adaptation: A conceptual framework. Tyndall Centre Working Paper No. 38.

Brown B, Chanakira R R, Chatiza K, Dhliwayo M, Dodman D, Masiiwa M, Muchadenyika D, Mugabe P and Zvigadza S (2012) Climate change, impact and vulnerability. London: International Institute for Environment and Development (IIED) Climate Change Working Paper No. 3.

Castree N (2014) The Anthropocene and environmental humanities: Extending the conversation. *Environmental Humanities*, Vol 5: 233-260.

Chambers R (1983) *Rural development: Putting the last first*. New York: Longman.

Chambers R (1989) Reversals, institutions and change. In: Chambers R, Pacey A and Thrupp L A (eds), *Farmers first. Farmer innovation and agricultural research*. London: Intermediate Technology Publications, 181-195.

Chimhowu A and Woodhouse P (2008) Communal tenure and rural poverty: Land transactions in Svosve communal area, Zimbabwe. *Development and Change* 39 (2): 285-308.

Clapham C (2012) From liberation movement to government: Past legacies and the challenge of transition in Africa. Discussion paper 8. Johannesburg: The Brenthurst Foundation.

Cousins B (2010) What is a 'smallholder'? Class-analytic perspectives on small-scale farming and agrarian reform in South Africa. *Institute of Poverty, Land and Agrarian Studies,* 24. University of the Western Cape, South Africa.

Cousins B (Undated) Key features of African land tenure. Unpublished.

Crane T A (2014) Bringing science and technology studies into agricultural anthropology: Technology development as cultural encounter between farmers and researchers. *The Journal of Culture and Agriculture,* 36 (1): 45-55.

Crate S A and Nuttal M (2009) Introduction: Anthropology and climate change. In: Crate S A and Nuttall M(eds), *Anthropology and climate change: From encounters to actions.* California: Left Coast Press, 9-38.

Critchley W, Cooke R, Jallow T, Lafleur S, Laman M, Njoroge J, Nyagah V and Obas V (1999) Harnessing local environmental knowledge in East Africa. Workshop Report No. 2. United Nations Development Programme (UNDP) and Swedish International Development Agency (SIDA)'s Regional Land Management Unit (RELMA).

Davis C L (2011) *Climate risk and vulnerability: A handbook for Southern Africa.* Council for Scientific and Industrial Research, Pretoria, South Africa.

Del Sesto D. 2013. *Shift points: Shift your thinking, change your life.* USA: Revell

Department of Geography and Environmental Studies (2010) Land use map of Zimbabwe showing location of Zvishavane. Harare: University of Zimbabwe.

DeWaal A (1989) *Famine that kills: Darfur, Sudan, 1984 – 1985.* Oxford: Clarendon Press.

DFID (UK Department for International Development) (2011) *Defining disaster resilience: A DFID approach paper.* London.

Dobie P, Barklund A and Hurni H (2001) Foreword. In: Mutunga K and Critchley W, Lameck P, Lwakuba A and Mburu C (eds), *Farmers' initiatives in land husbandry: Promising technologies for the drier areas of East Africa*. Nairobi: Regional Land Management Unit (RELMA), Swedish International Development Cooperation Agency (Sida). (RELMA Technical Report Series: 27), viii-ix.

Elwell H A (1985) An assessment of soil erosion in Zimbabwe. *Zimbabwe Science News* 19 (3/4): 27 – 31.

Eriksen T H (2002) *What is Anthropology?* London: Pluto Press.

Everson C, Everson T M, Modi A T, Csiwila D, Fanadzo M, Naiken V, Auerbach R M B, Moodley M, Mtshali S M and Dladla R (2011) *Sustainable techniques and practices for water harvesting and conservation and their effective application in resource poor agricultural production through participatory adaptive research*: report to the Water Research Commission. Gezina (South Africa): Water Research Commission. Available at:
www.wrc.org.za/Knowledge%20Hub%20Documents/Researc h%20Reports/1465-1-11.pdf.

Farmer-led Documentation (FLD) (2012) Introducing Prolinnova. Available at:
www.prolinnova.net/sites/default/files/documents/resources/ publications/2012/FLD_booklet_2012/chapter_1.pdf.

Feirman S (1990) *Peasant intellectuals: Anthropology and history in Tanzania*. Madison: The University of Wisconsin Press.

Fetien A, Gandarillas E, Shrestha P, Waters-Bayer A and Wongtschowski M (2009) Recognising and enhancing local innovation in managing agricultural biodiversity. Prolinnova Working Paper 29.

Fetterson D M (1998) *Ethnography. Step by step*. Los Angeles: Sage Publications.

Folke C, Carpenter S R, Walker B, Scheffer M, Chapin T and Rockstrom J (2010) Resilience thinking: Integrating resilience, adaptability and transformability. *Ecology and Society* 15 (4): 1-9.

Fuller B (2008) Buckminster quotes.
Availableat:www.goodreads.com/author/quotes/11515303.R_ Buckminster_Fuller.

Gann L (1964) *A history of Southern Rhodesia: Early days to 1934.* New York: Humanities Press.

Geertz C (1973) *The interpretation of cultures: Selected topics.* New York: Basic Books.

Giblin J and Maddox G (1996) Introduction. In: Maddox G, Giblin J and Kimambo I N (eds), *Custodians of the land: Ecology and culture in the history of Tanzania.* London: James Currey Ltd, 15-18.

Gijsbers G W (2009) *Agricultural innovation in Asia: Drivers, paradigms and performance.* PhD thesis, Erasmus University Rotterdam, Netherlands.

Gitz V and Meybeck A (2012) Risks, vulnerabilities and resilience in a context of climate change. In: Meybeck A, Lankosi J, Redfern S, Azzu A and Gitz V (eds), *Building resilience for adaptation to climate change in the agriculture sector.* Proceedings of a joint Food and Agriculture Organisation/Organisation for Economic Cooperation and Development(FAO/OECD) workshop. FAO, 19-36.

Global Alliance for Climate-Smart Agriculture Action Plan (2014) Available at: www.un.org./climate change/summit/wp-content/uploads/sites/2/2014/09/AGRICULTURE-ActionPlan.pdf.

Gobo G (2008) *Doing ethnography.* Los Angeles: Sage Publications.

Goldman A (1993) Agricultural innovation in three areas of Kenya. Neo-Boserupian theories and regional characterisation. *African Development* 69 (1): 44-71.

Government of Zimbabwe (2009) National Environmental Policy. Harare.

Green L (2008a) Anthropologies of knowledge and South Africa's indigenous knowledge systems policy. *Anthropology Southern Africa,* 31 (1-2): 48-57.

Green L (2008b) 'Indigenous knowledge' and 'Science': Reframing the debate on knowledge diversity. *Archaeologies: Journal of the World Archaeological Congress* 4(1):144-163.

Hall A J Suliaman V, Clark N G and Yoganand B (2003) From measuring impact to learning institutional lessons: an innovation systems perspective on improving the management of international agricultural research. *Agricultural Systems* 78:213-241.

Hall A J, Mytelka L and Oyelaran-Oyeyika B (2005) Innovation systems: Implications for agricultural policy and practice. *International Learning and Change Initiative* (ILAC) Brief 2.

Hall A J (2009) Challenges to strengthening agricultural innovation systems: Where do we go from here? In: Scoones I Chambers R and Thompson J, *Farmers first revisited: Innovation for agricultural research and development.* Warwickshire: Practical Action Publishing Ltd, 30-38.

Hall A J, Clark N and Frost A (2010) Bottom-up, bottom-lines: Development-relevant enterprises in East Africa and their significance for agricultural innovation. United Nations University – Maastricht Economic and Social Research Institute on Innovation and Technology(UNU-MERIT), The Netherlands. Working Paper 2010-42.

Hammersley M (1998) *Reading ethnographic research. A critical guide.* New York: Addison Wesley Longman Limited.

Hans-Martin F (2007) Vulnerability: A generally applicable conceptual framework for climate change research. *Global Environmental Change* 17: 155-167.

Harrison P (1987) *The greening of Africa. Breaking through in the battle for land and food.* London: Earthscan.

Hastrup K (2009) *The question of resilience. Social responses to climate change.* Copenhagen: The Royal Danish Academy of Sciences and Letters.

Hastrup K (2013) Anthropological contributions to the study of climate: Past, present, future. Wiley Interdisciplinary Reviews Climate Change (*WIREs Clim Change*), 4:269-281/doi:10.1002/wcc.219.z.

Hiller D and G Castillo (2013) No accident: Resilience and the inequality of risk. Briefing Paper 172. Boston, MA, US: Oxfam.

Holy Bible (2008) New Revised Version. Massachusetts: Hendrickson Publishers.

Howden S M, Crimp S and Nelson R (2010) Australian agriculture in a climate of change. In: Jubb I, Holper P, Cai W (eds), *Managing climate change: Papers from the Greenhouse 2009 Conference*. Melbourne: CSIRO Publishing, 101-111.

Howe C and A Pandian (2016) Introduction: Lexicon for an Anthropocene yet unseen. Theorising the contemporary, Cultural Anthropology website, January 21. Available at: https://culanth.org/fieldsights/788-introduction-lexicon-for-an-anthropocene-yet-unseen

Hulme M (2009) *Why we disagree about climate change: Understanding controversy, inaction and opportunity*. UK: Cambridge University Press.

Hulme M (2011) Reducing the future to climate: A story of climate determinism and reductionism. *Osiris* 26 (1): 245–266.

Iliffe, J (1990) *Famine in Zimbabwe 1890-1960*. Gweru: Mambo Press.

Intergovernmental Panel on Climate Change Third Assessment Report (IPCC TAR) (2001) Climate change 2001: Impacts, adaptation and vulnerability. *IPCC Third Assessment Report*. Cambridge: Cambridge University Press.

Intergovernmental Panel on Climate Change (IPCC) (2007) The physical science basis. *Fourth Assessment Report of the IPCC*. Cambridge and New York: Cambridge University Press.

Intergovernmental Panel on Climate Change (IPCC) (2014) Climate Change. Fifth Assessment Report (AR5). (Core writing team and Pachauri R K, Meyer L A) (eds). Geneva Switzerland: IPCC.

International Food Policy Research Institute (IFPRI) (2002) Green Revolution. Washington DC: IFPRI.

Jayne T S, Chisvo M, Rukuni M and Masangaise P (2006) Zimbabwe's food insecurity paradox: Hunger amid potential. In: Rukuni M, Tawonezvi P, Eicher C, Munyuki-Hungwe M and Matodi P (eds), *Zimbabwe's agricultural revolution revisited*. Harare: University of Zimbabwe Publications, 525-542.

Jaser Z (2016) Post-truth leaders are all about their followers. *The Conversation*, 24 November.

Juma C (2011) *The new harvest: Agricultural innovation in Africa*. New York: Oxford University Press.

Kabwe G, Bigsby H and Cullen R (2009) Factors influencing adoption of agroforestry among smallholder farmers in Zambia.

Paper presented at the 2009 New Zealand Agricultural and Resource Economics Society (NZARES) Tahuma Conference Centre – Nelson, New Zealand. August 27 – 28.

Kay R C (2009) Adaptation by ribbon cutting. *Tiempo*, 73, 11–15. Available at: http://www.tiempocyberclimate.org/newswatch/xp_comment 090622.htm.

Keohane G (2015) MLK, civil rights and the fierce urgency of the now. *Time Magazine*, 19 January.

Klein N (2013) Green groups may be more damaging than climate change deniers. *The Gurdian*. Thursday, September 10th. Available at: www.thegurdian.com/environment/2013/sep/10/naomi-klein-green-groups-climate-deniers.

Kopytoff I (1987) *The african frontier*. Bloomington: Indiana University Press.

Lakoff G and Johnson M (1980) *Metaphors we live by*. Chicago: The University of Chicago Press.

Lancaster B (1999) The man who farms water. *Aridlands Newsletter*. No. 46, Fall/Winter. Available at: ag.arizona.edu/oals/ALN/aln46/Lancaster.html.

Lancaster B (2008) Case study: Drought resistant farming in Africa. *Ecologist*, November. Available at: www.theecologist.org/campaigning/food_and_gardening/3602 57/case_study_drought_resistant_farming_in_africa.html.

Law J (2004) *After method mess in social science research*. London: Routledge.

Leach M, Scoones I and Stirling I (2010) *Dynamic sustainabilities: Technology, environment and justice*. London: Earthscan.

Letty B and Bell M (2012) Grassroots innovation: No support without solid evidence. Science and Development Network (*SciDev.net*). Available at: http://www.scidev.net/en/science-and-innovation-policy/opinions/grassroots-innovation-no-support-without-solid-evidence.html.

Levina E and Tirpark D (2006) *Adaptation to climate change: Key terms.* Paris: Organisation for Economic Cooperation and Development (OECD).

Li X Y, Cook S, Gebelle G T and Burch W R (2000) Rainwater harvesting in agriculture: An integrated system for water management on rainfed land in China's semiarid areas. A journal of the Human Environment (*Ambio*) 29 (8): 477-483.

Little P D and McPeak J G (2014) Pastoralism and resilience south of the Sahara. 2020 Conference Brief 9. Washington: International Food Policy Research Institute.

Lliffe J (1990) *Famine in Zimbabwe 1890 – 1960.* Gweru: Mambo Press.

Maathai W M (2006) *Unbowed: A memoir.* London: Arrow Books.

Mabeza C (2013) Metaphors for climate adaptation from Zimbabwe: Zephaniah Phiri Maseko and the marriage of water and soil. In: Green L (ed), *Contested ecologies, dialogues in the south on nature and knowledge.* Cape Town: HSRC Press, 126-137.

Mabeza C, T Pesanayi and C Luphele. Forthcoming. Business unusual: Towards transformational adaptation.

Magadza C D H (2000) Climate change impacts and human settlements in Africa: Prospects for adaptations. Proceedings of United Nations Environment Programme (UNEP) Conference on climate change and human settlements, San Jose, Costa Rica, 29 March-1 April.

Maldonado-Toress N (2008) *Against war: Views from the underside of modernity.* Durham: Duke University Press.

Mangoma J F (2011) *The effects on local livelihoods of wetland development scheme in a Zimbabwe village: An ethnographic study.* PhD thesis, University of Cape Town. South Africa.

Manyanga M, Pikirayi I and Ndoro W (2000) Coping with dryland environments: Preliminary results from Mapungubwe and Zimbabwe phase sites in the Mateke Hills, South-Eastern Zimbabwe. *Goodwin Series* 8:69-77.

Manyanga M (2006) *Resilient landscapes. Socio-environmental dynamics in the Shashi-Limpopo basin, southern Zimbabwe.AD 800 to the present.* PhD thesis, Uppsala University, Sweden.

Mapedza E and Mandondo A (2002) Co-management in the Mafungautsi state forest area of Zimbabwe: What stake for local communities? *Environmental Governance in Africa* Working Paper 5. Washington DC: World Resources Institute.

Maravanyika-Mutimukuru T (2010) *Can we learn our way to sustainable management? Adaptive collaborative management in Mafungsbutsi state forest, Zimbabwe.* PhD thesis, Wageningen University, Netherlands.

Marwick A and Emsley C (1989) Introduction. In: Emsley C, Marwick A and Simpson W(eds), *War, peace and social change in twentieth century Europe.* United Kingdom: Milton Keynes, 1-23.

Mawere A and Wilson K (1995) Socio-religious movements, the state and community change: Some reflections on the Ambuya Juliana cult of southern Zimbabwe. *Journal of Religion in Africa* 25. 252-287.

Mawere M and Mabeza C M (2015) Sheep in sheep's clothing or wolves in sheep's clothing? Interventions by non-state actors in a changing climatic environment in rural Zimbabwe. In: Mawere M and Awuah-Nyamekye S (eds), *Between rhetoric and reality: The state and use of indigenous knowledge in post-colonial Africa.* Bamenda: Langaa RPCIG, 167-180.

Mawire G (2013) *Hurudza.* Available at:
 http://www.facebook.com/ntceu/posts/5464850655441052 (unpublished).

Mehretu A and Mutambirwa C C (2006) Social poverty profile of rural agricultural areas. In: Rukuni M, Tawonezvi T, Eicher C, Munyuki-Hungwe M and Matodi P (eds), *Zimbabwe's agricultural revolution revisited.* Harare: University of Zimbabwe Publications, 119-140.

Meybeck A, Lankosi J, Redfern S, Azzu N and Gitz V (2012) Building resilience for adaptation to climate change in the agriculture sector. Proceedings of a Joint Food and Agriculture/Organisation for Economic Cooperation and Development (FAO/OECD) Workshop 23 – 24 April. FAO, Rome.

Mhlanga C (2014) Of pentecostalism, individualism, selfishness. *The Daily News*, 20 January. Available at: http://www.dailynews.co.zw/articles/2014/01/20/of-pentecostalism-individualism-selfishness.

Ministry of Environment and Natural Resources Management (1998) Initial communication on climate change. Zimbabwe.

Ministry of Environment and Natural Resources Management (2009) National Environmental Policy and Strategies. Zimbabwe.

Ministry of Environment and Natural Resources Management, Climate change office, Zimbabwe (Undated) Climate variability and change in Zimbabwe.

Ministry of Environment and Natural Resources Management, Climate change office, Zimbabwe (Undated) Fact sheet.

Mlambo A S (2009) From the second world war to UDI, 1940-1965. In: Raftopoulos B and Mlambo A S (eds), *Becoming Zimbabwe: A history from pre-colonial period to 2008*. Harare: Weaver Press, 75-114.

Moberg F/Stockholm Resilience Centre (2014) *What is resilience? An introduction to socio-ecological research*.Stockholm University: Stockholm Resilience Centre.

Morton J F (2007) The impact of climate change on smallholder and subsistence agriculture. *The National Academy of Sciences of the USA*104 (50): 19680-19685.

Moser S C and Ekstrom J A (2010) A framework to diagonise barriers to climate change adaptation. *Proceedings of the National Academy of Sciences of the United States of America,* 107(51): 22026-22031.

Moyana H V (1984) *The political economy of land in Zimbabwe.* Gweru:Mambo Press.

Msipa C (2014) Interview with Dr Msipa, 24 January. Gweru, Zimbabwe.

Mtisi J, Nyakudya M and Barnes T (2009) Social and economic developments during the UDI Period. In: Raftopoulos B and Mlambo A, *Becoming Zimbabwe: A history from pre-colonial period to 2008*.Harare: Weaver Press, 115-140.

Muir J (2014) Islamic Sate: Radical shifts needed to combat threat. British Broadcasting Cooperation (*BBC*) *News*, 22 August.

182

Available at:http://www.bbc.com/news/world-middle-east-28900096.

Muir-Leresche K (2006) Agriculture in Zimbabwe. In: Rukuni M, Tawonezvi T, Eicher C, Munyuki-Hungwe M and Matodi P. *Zimbabwe's agricultural revolution revisited.* Harare: University of Zimbabwe Publications, 99-108.

Murwira A, Masocha M, Gwitira I, Shekede M D, Manatsa D and Mugandani R (2012) Vulnerability and adaptation assessment.35-54. In: Zhakata W (ed), *Zimbabwe Second National Communication to the United Nations Framework Convention on Climate Change.* Harare:Sable Press (Pvt) Ltd, 35-54

Murwira A, D Manatsa, Mushonjowa E, R Mugandani, C P Mubaya and I Gwitira (Forthcoming) Third National Communication to the UNFCCC. Zimbabwe climate change vulnerability and adaptation assessment.

Mushongah J (2009) *Rethinking vulnerability: Livelihood change in southern Zimbabwe 1986-2006.* PhD thesis, University of Sussex,United Kingdom.

Musiyiwa M (Forthcoming) Shona as a land-based nature culture: A study of the (re) construction of the Shona land mythology in popular songs. In:Moolla F (ed),*African natures-cultures: Environment and animals in African cultural forms.* Johannesburg: Wits University Press.

Mutimukuru-Maravanyika T (2010) *Can we learn our way to sustainable management? Adaptive collaborative management in Mapfungabutsi state forest, Zimbabwe.* PhD thesis, Wageningen University,Netherlands.

Mutunga K and Critchley W (2010) *Farmers' initiatives in land husbandry: Promising technologies for the drier areas of East Africa.* Regional Land Management Unit (RELMA) Technical Report Series 27.

Naess A (1973) The shallow and the deep, long range ecology movements: A summary. *Inquiry* (Oslo) 16: 95-100.

National Museums and Monuments of Zimbabwe. (Undated) Nyanga hill terracing.

New Revised Standard Version Bible (1989) National Council of the Churches of Christ in the United States of America.

Nguthi F N (2007) *Adoption of agricultural innovations by smallholder farmers in the context of HIV/AIDS: The case of tissue-cultured banana in Kenya*. PhD thesis, Wageningen University, Netherlands.

Nicol A, Langan S, Victor M and Gonsalves J (2015) *Water-smart agriculture in East Africa*. CGAIR Research Programme on Water, Land and Ecosystems (WLE); Kampala, Uganda. Cooperative for assistance and Relief Everywhere (CARE). doi: 10.5337/2015.203.

Nkomo J (2001) *Nkomo, the story of my life*. Harare: Sapes Books.

Nyambara P (2001) 'Immigrants, "Traditional" leaders and the Rhodesian State: the Power of "Communal" Land Tenure and the Politics of Land Acquisition in Gokwe, Zimbabwe, 1963-1979', *Journal of Southern African Studies* 27 (4): 771-791.

Nyamnjoh F (2002) 'A child is one person's only in the womb', Domestication agency and subjectivity in the Cameroonian Grassfields. In: Werbner R (ed), *Postcolonial subjectivities in Africa*.London: Zed Books, 109-138.

Nyamnjoh F (2013a) Fiction and reality of mobility in Africa. *Citizenship Studies* 17 (6-7): 653-680.

Nyamnjoh F (2013b) Interview I held with him in his office on 5 November.

O'Brien K, Quinlan T and Ziervogel G (2009) Vulnerability interventions in the context of multiple stressors: Lessons from Southern Africa Vulnerability Initiative (SAVI). *Environmental Science and Policy* 12 (1): 23-32.

O'Brien K (2012) Global environmental change II: From adaptation to deliberate transformation. *Progress in Human Geography* 36 (5): 667-676.

O'Brien K (2013) The courage to change, adaptation from the inside-out. In: Moser S C and Boykoff M T (eds), *Successful adaptation to climate change, linking science and policy in a rapidly changing world*. Oxfordshire, United Kingdom: Routledge, 307-319.

O'Brien K. 2013.What's the problem? Putting global environmental change into perspective (draft). *World Social Science Report*

Orlove B S, Chiang J C H and Crane M A (2002) Ethnoclimatology in the Andes: A cross-disciplinary study uncovers the scientific basis for the scheme Andean potato farmers traditionally use to predict the coming rains. *American Scientist* 90: 428-435.

Ortony A (1993) Metaphor, language, and thought. In: Ortony A (ed), *Metaphor and thought*. Cambridge, England: Cambridge University Press, 1-16.

Oxfam GB Report. Undated. *Mr Zephaniah Phiri's book of life*. Unpublished.

Palmer R (1977) *Land and racial domination in Rhodesia*. Berkeley: University of California Press

Pandian A (2015) *Reel World: An Anthropology of Creation*. Durham, N.C.: Duke University Press.

Park S E, Marshall N A, Jakku E, Dowd A M, Howden S M, Mendham E and Fleming A (2012) Informing adaptation responses to climate change through theories of transformation. *Global Environmental Change* 22 (1): 115 -126.

Pelling M (2011) *Adaptation to climate change: From resilience to transformation*. Abingdon: Routledge.

Pelling M, O'Brien K and Matyas D (2014) Adaptation and transformation. *Climate Change*, 1-15. doi 10.1007/s10584-014-1303-0.

Pereira L (2014) Orphan crop innovation for transformation in the food system. An African Climate Development Initiative (ACDI) Seminar Series paper presented on 27 February.

Phimister I (1988) *An economic and social history of Zimbabwe 1896-1948: Capital accumulation and class struggle*. New York: Longman.

Pielke R A Jr (2009) Rethinking the role of adaptation in climate policy. In: Schipper E L F and Burton I (eds), *Adaptation to climate change*. London: Earthscan, 345-358.

Piot C (1999) *Remotely global*. Chicago: The University of Chicago Press.

Pisano U (2012) Resilience and sustainable development: Theory of resilience, systems thinking and adaptive governance. European Sustainable Development Network (*ESDN*) *Quarterly Report No 26*. Vienna: European Sustainable Development Network.

Preiser R (2014) SAPECS Winter school, an introduction to complexity presentation. Nelson Mandela Metropolitan University: South Africa.

Ramos J C (2016) *The seventh sense: Power, fortune and survival in the age of networks*. USA: Little, Brown and Company.

Ranger T (2002) Listening, Books from Zimbabwe. *Journal of Southern African Studies* 28 (1): 199-206.

Rankine C (2015) The meaning of Serena Williams. *The New York Times Magazine*, 15 August. Available at: http://www.nytimes.com/2015/08/30/magazine/the-meaning-of-serena.html?_1=0.

Reddy V T S, Hall A and Sulaiman R (2010) New organisational and institutional vehicles for managing innovation in South Asia: Opportunities for using research for technical change and social gain. Unu-Merit Working Paper Series 054.

Reid P and Vogel C (2006) Living and responding to multiple stressors in South Africa – Glimpses from KwaZulu-Natal. *Global Environmental Change* 16 (2): 195-206.

Rhoades R E (1989) The role of farmers in the creation of agricultural technology. In: Chambers R, Pacey A and Thrupp L A (eds), *Farmers first: Farmer innovation and agricultural research*, London: Intermediate Technology Publications, 3-9.

Ribot J (2011) Vulnerability before adaptation: Toward transformative climate action. *Global Environment Change* 21: 1160-1162.

Ribot J (2014) Cause and response: Vulnerability and climate in the Anthropocene. *The Journal of Peasant Studies*, DOI: 10.1880/03066150.2014.89491. Available at: http://dx.doi.org/10.1080/03066150.2014.894911.

Rickards L and Howden S M (2012) Transformational adaptation: Agriculture and climate change. *Crop and Pasture Science* 63: 240–250.

Rickards L (2013) Transformation is adaptation. *Nature Climate Change* 3: 690.

RobinsS L (1995) *Close encounters at the 'development' interface: Local résistance, state power and the politics of land-use planning in Matebeleland,*

Zimbabwe. PhD thesis, Columbia University, United States of America.

Rodima-Taylor D, Olwig M F and Chhetri N (2012) Adaptation as innovation, innovation as adaptation: An institutional approach to climate change. *Applied Geography* 33: 107-112.

Rogers E M (1976) New product adoption and diffusion. *Journal of Consumer Research* 2: 290-301.

Rogers E M (1983) *Diffusion of innovations*. New York: The Free Press.

Rogers E M (2003) *Diffusion of Innovations* (5th edition). New York: The Free Press.

Roling N (1988) *Extension Science: Information systems in agricultural development*. Cambridge, United Kingdom: Cambridge University Press.

Roling N (2009) Conceptual and methodological developments in innovation. In: Sanginga P C, Waters-Bayer A, Kaaria J, Njuk J and Wettasinha C (eds), *Innovation Africa. Enriching farmers' livelihoods*. UK: Earthscan, 9-34.

Roncoli C, Crane T and Orlove B S (2009) Fielding climate change in Cultural Anthropology. In: Crate S A and Nuttall M(eds), *Anthropology and climate change: From encounters to actions*. California: Left Coast Press, 87-115.

Rowland J K (2008) The fringe benefits of failure, and the importance of imagination. Available at: news.harvard.edu/gazette/story/2008/06/text-of-j-k-rowling-speech/

Rukuni M (2006) The evolution of agricultural policy: 1890-1990. In: Rukuni M, Tawonezvi T, Eicher C, Munyuki-Hungwe M and Matodi P (eds), *Zimbabwe's agricultural revolution revisited*. Harare: Sable Press Private Limited, 29-61.

Sanginga P C, Waters-Bayer A, Kaaria J, Njuk J and Wettasinha C (eds) (2009) *Innovation Africa. Enriching farmers' livelihoods*.UK: Earthscan.

Schipper E L F and Burton I (2009) Understanding adaptation: Origins, concepts, practice and policy. In: Schipper E L F and Burton I (eds), *Adaptation to climate change*. London: Earthscan, 1-10.

Scoones I (1988) *Learning about wealth: An example from Zimbabwe.* Centre for Environmental Technology. London: London Imperial College.

Scoones I (2004) On the nomination of Mr Phiri Maseko for the 2004/2005 King Baudouin International Development Prize. In: Wilson K, *Zephaniah Phiri's book of life.* Unpublished.

Scoones I and Thompson J (2009) *Farmer first revisited: Innovation for agricultural research and development.* Warwickshire: Practical Action.

Scoones I (2014) Recognising farmer innovation: the launch of the Phiri award. Zimbabwe land. Available at: https://zimbabweland.wordpress.com/2014/10/06/recognising-farmer-innovation-the-launch-of-the-phiri-award/.

Secretary for Internal Affairs (1963) *Report of the Secretary for Internal Affairs for the year (1963)* Salisbury (Harare): Government Printers.

Shackleton S E and project team (2013a) Barriers to local level climate change adaptation amongst poor communities in Africa: What do we know and what does this mean for adaptation? Findings from literature review WUN Project on Limits to Adaptation. A paper presented at the *Southern African Adaptation Colloquium*, 25 – 26 November, Cape Town, South Africa.

Shackleton S E and project team (2013b) Environmental Science, Rhodes University. Exploring the factors that affect local people's ability to respond to multiple stressors in two communal areas of the Eastern Cape, South Africa. A paper presented at *the Southern African Adaptation Colloquium* 25 – 26 November. Cape Town, South Africa.

Shackleton S E and Shackleton C M (2012) Linking poverty, HIV/AIDS and climate change to human and ecosystem vulnerability in southern Africa: consequences for livelihoods and sustainable ecosystem management. *International Journal of Sustainable Development & World Ecology*, 19:3, 275-286, DOI:10.1080/13504509.2011.641039.

Sieden L S (2012) Buckminster Fuller's synergy solutions for today. Available at: www.buckyfullernow.com/blog---a-fuller-view---wwbs-what-would bucky-say/archives/04-2012/2.

Simonsen S H, Biggs R, Schluter M (2014) *Applying resilience thinking. Seven principles for building resilience in social-ecological systems.* Stockholm: Stockholm University.

Skrydstrup M (2009) Planetary resilience: Codes, climates and cosmo-science in Copenhagen. In: Hastrup K (ed), *The question of resilience, responses to climate change.* Copenhagen: The Royal Danish Academy of Sciences and Letters, 336-358.

Sledzik K (2013). Schumpeter's view on innovation and entrepreneurship. In: Hittmer S (ed), *University of Zilina and Institute of Management.* Slovakia: University of Zilina.

Sluka J A and Robben A C G M (2007) Fieldwork in cultural anthropology: An introduction. In: Robben A C G M and Sluka J A (eds), *Ethnographic fieldwork.* Oxford: Blackwell, 1-28.

Smith B (2014) World Cup: How Costa Rica's 'bulls' shocked the world. British Broadcasting Cooperation(*BBC*) *Sport.* Available at: www.bbc.com/sport/0/football/27951700.

Smith D (2014) Chido Govera: transforming lives in rural Africa by growing mushrooms. *The Observer*, 16 August, United Kingdom. Available at:
www.thegurdian.com/lifestyle/2014/aug/16/chido-govera-mushrooms-zimbabwe-changing-lives.

Smithers J and Smit B (2009) Human adaptation to climatic variability and change. In:Schipper E L F and Burton I (eds),*Adaptation to climate change.* London: Earthscan, 15-33.

Soper R (2002) Nyanga: Ancient fields, settlements and agricultural history in Zimbabwe. Memoirs of the British Institute in Eastern Africa Number 16. Nairobi: The British Institute in Eastern Africa.

Soper R (2006) *The terrace builders of Nyanga.* Harare: Weaver Press.

Stafford-Smith M, Horrocks L, Harvey A and Hamilton C (2011) Rethinking adaptation for a 4-degree Celsius world. *Philosophical transactions of the Royal Society*, 369: 196-216.

Steimetz K (2014) Recycle, reuse, reprofit. Startups are trying to make money selling your unwanted stuff. *Time Magazine*, August 4, 11.

Stoknes P E (2015) What we think about when we try not to think about global warming. *The New Psychology of climate action*. Chelsea Green Publishing. Available at: http://www.chelseagreen.com/what-we-think-about-when-we-try-not-to-think-about-global-warming.

Stromberg J (2013) What is the anthropocene and are we in it? *Smithsonian Magazine*, January. Available at: http://www.smithsonianmag.com/science-nature/what-is-the-anthropocene-and-are-we-in-it-164801414/? no-ist.

St. Augustine's press (2016) They had learned nothing and forgotten nothing. Available at: www.staugustine.net/blogs/rectify-names-a-blog-on-publishing/e2809cthey-had-learned-nothing-and-forgotten-nothinge2809d-march-11-2013/

Sugunan V V (2007) Fisheries management of small water bodies in seven countries in Africa, Asia and Latin America. Rome: FAO.

Sullivan K and J Y Smith (2016) Fidel Castro, revolutionary leader who remade Cuba as a socialist state, dies at 90. *The Washington Post*, 26 November.

Tekere E Z (2007) *A lifetime of struggle*. Harare: Sapes Books.

Tepperman J (2016) *The fix: How nations survive and thrive in a world in decline*. Canada: Tim Duggan Books.

The Economist (2016) Post-truth politics: Art of the lie. 10 September. Available at: http://www.economist.com/news/leaders/21706525-politicians-have-always-lied-does-it-matter-if-they-leave-truth-behind-entirely-art

The Global Alliance for Climate-Smart Agriculture (2014) Available at: http://www.fao.org/climate -smart-agriculture/85725/en/.

Thomas D S G, Osbahr H, Twyman C, Adger N and Hewitson B (2005) Adaptive: Adaptations to climate change amongst natural resource-dependent societies in the developing world: across the Southern African climate gradient. Tyndall Centre Technical Report No.35, University of East Anglia., UK.

Thomas D S G, Twyman C, Osbahr H and Hewitson B (2007) Adaptation to climate and variability: farmer responses to intra-seasonal precipitation trends in southern Africa: farmer responses to intra-seasonal precipitation trends. *Climatic Change* 83 (3):301-322.

Tiffen M (1996) Land and capital: Blind spots in the study of the "resource poor" farmer. In: Leach M and Mearns R (eds), *The lie of the land: Challenging received wisdom on the African environment.* Oxford: James Currey, 168-185.

Tschakert P and Dietrich K N (2010) Anticipatory learning for climate change adaptation and resilience. *Ecology and Society* 15 (2): 11-34.

Tschakert P, van Oort B, St Clair A L and LaMadrid A (2014) Inequality and transformation analyses: A complimentary lens for addressing vulnerability to climate change. *Climate and Development* 5(4): 340-350.

United Nations Development Programme (UNDP) (2014) Sustaining human progress: Reducing vulnerabilities and building resilience. *Human Development Programme.* New York.

Vamsidhar R, Hall A and Sulaiman R (2012) Locating research in agricultural innovation trajectories: Evidence and implications from empirical cases from south Asia. *Science and Public Policy,* 1-15.

Van Onselen C (1976) *Chibharo: African mine labour in Southern Rhodesia, 1900-1933.* London: Pluto Press.

Van Onselen C (1996) *The seed is mine; the life of Kas Maine, a South African sharecropper 1894-1985.* Cape Town: David Philip.

Veteto J R and Crane T A (2014) Tending the field: Special issue on agricultural anthropology and Robert E Rhoades. *The Journal of Culture and Agriculture,* 36 (2): 1-3.

Vincent V and Thomas R G (1960) An Agricultural Survey of Southern Rhodesia. Part 1: Agro – ecological Survey. Salisbury: Government Printer.

Virde M (2016) "Life hacking": Building resilience through innovation. Accessed at:
www.braced.org./news/i/?id=302ac134-09a5-429c-8d0c-1563217f94a6

Vogel C, Moser S C, Kasperson R E and Dabelko G D (2007) Linking vulnerability, adaptation, and resilience science to practice: pathways, players, and partnerships. *Global Environmental Change* 17 (3-4): 349-64.

Walker B and Salt D (2006) *Resilience thinking: Sustaining ecosystems and people in a changing world.* Washington: Island Press.

Wall M (2014) Innovate or die: The stark message for big business. British Broadcasting Cooperation (*BBC*) *News*. 4 September. Available at: http://www.bbc.com/news/business-28865268.

Waters-Bayer A and van Veldhuizen L (2004) Promoting local innovation: Enhancing IK dynamics and links with scientific knowledge. I K Notes No. 76 January.

Waters-Bayer A (Undated) Climate-resilient agriculture (CRA) and agricultural innovation systems (AIS). AIS Workshop Working Group Session 3.

Waters-Bayer A and van Veldhuizen L (Undated) The International Support Team for Prolinnova: the International Institute for Rural Reconstruction (IIRR) in the Philippines, the Swiss Centre for Agricultural Extension (LBL), the Centre for International Cooperation of the Free University of Amsterdam and ETC and Ecoculture in the Netherlands.

Weisser F, Bolling M, Doevenspeck M and Mahn-Muller D (2014) Translating the 'adaptation to climate change' paradigm: The politics of a travelling idea in Africa. *The Geographical Journal* 180(2): 111-119.

Whitlow J R (1988) *Land degradation in Zimbabwe: A geographic study.* Harare: Natural Resources Board.

Wilson J (2012) Phiri award for food sovereignty (unpublished).

Wilson K B (2010) *Overview of Zephaniah Phiri's book of life, on the occasion of Mr Phiri Maseko's lifetime achievement award.* August 24th. Harare: University of Zimbabwe.

Wilson K B (2010) *Zephaniah Phiri's book of life.* Unpublished.

Wilson K (Undated) A lifetime's records of innovation and experimentation: Mr Phiri's water, soil and landscape management principles as exemplified on his own land. Unpublished.

Wisner B, Blaikie P, Cannon T and Davis I (2003) *At risk, natural hazards, people's vulnerability and disasters.* London: Routledge.

Witoshynsky M (2000) *The water harvester: Episodes from the inspired life of Zephaniah Phiri.* Harare: Weaver Press.

Woddis J (1960) *The roots of revolt.* London: Lawrence and Wishart.

World Bank (2006) *Enhancing agricultural innovation: How to go beyond the strengthening of research systems.* Economic Sector Work Report. Washington DC: World Bank.

World Bank Independent Evaluation Group (IEG) (2012) *Adapting to climate change: Assessing the World Bank Group Experience Phase 111.* Washington D C: World Bank.

Zhakata W (ed) (2012) *Zimbabwe Second National Communication to the United Nations Framework Convention on Climate Change.* Harare: Sable Press (Pvt) Ltd.

Ziervogel G (2002) *Seasonal climate forecast applications: A case study of smallholder farmers in Lesotho.* PhD Thesis, University of Oxford, UK.

Ziervogel G and Calder R (2003) Climate variability and rural livelihoods: Assessing the impact of seasonal climate forecasts in Lesotho. *Area* 35 (4): 403–417.

Ziervogel G, Bharwani S and Downing T E (2006) Adapting to climate variability. Pumpkins, people and policy. *Natural Resources Forum* 30: 294-305.

Ziervogel G and Taylor A (2008) Feeling stressed: Integrating climate adaptation with other priorities in South Africa. *Environment* 50(2): 32-41.

Ziervogel G and Zermogolio F (2009) Climate change scenarios and the development of adaptation strategies in Africa: Challenges and opportunities. *Climate Research* 40 (2-3): 133-146.

Ziervogel G and Ericksen P (2010) Adapting to change to sustain food security. Wiley Interdisciplinary Reviews (*WIREs*) *Climate Change* 1: 525-540.

Ziervogel G (2015) What is adaptation/climate adaptation/climate change adaptation? The Adaptation Network. Available at: www.adaptationnetwork.org.za/concepts/.

Zimbabwe Meteorological Services Department (Undated)Zvishavane rainfall graph from 1980 to 2014.

Zimbabwe National Statistics Agency (2012) Census: Preliminary Report. Available at: http://www.zimstat.co.zw/.

Zvishavane Water Project Report (1987) (Unpublished).

Zvishavane Water Project Report (1990) (Unpublished).

Zvishavane Water Project Progress Report (1992) (Unpublished).

Zvishavane Water Project Progress Report (1995) (Unpublished).

Zvishavane Water Project Green Business and Environmental Care in Chivi and Zvishavane Districts Proposal (2012) (Unpublished).

Zvishavane Water Project (2012) PRP Final Report, 27 August (Unpublished).

Glossary

Anemaoko haatsvi nenyemba	you have to be creative using available resources
Bamba zonkes	grab alls
Chikorokoza	gold panning
Chinamwe	clay soils
Chinamwe chiri mumusoro	soils exist in the mind
Chipani	span oxen for ploughing
Chisi	a sacred day
Dambo	Wetland
Dekete	Wetland
Dhiga udye	dig and eat
Dhiga ufe	dig and die
Dhiga ufe nekuguta	dig and 'die' of satisfaction from eating too much food
Divisi	a crop yield enhancing charm
Dube	Zebra
Dzivaguru	sacred pond
Goho guru	high yields
Gombo	virgin land
Gombo iri ndarida ndoririma chete	I will propose to this woman
Gora raipedza huku rakaurawa nemupfuuri	solutions can be found among foreigners
Hurudza	enterprising farmer
Hurudza yakafa nenzara	persistence in excellence
Husimbe	Laziness
Hazviparari izvi	The project will not collapse
Injiva	Zimbabweans working abroad perceived to be rich
Kohwa mvura kohwa pakuru	harvest water, harvest a bumper crop
Korokoza	gold panner

Kuronzera	to loan out livestock
Kushuzha	procuring grain in times of scarcity
Kutema ugariri	working for in-laws in lieu of paying bride wealth
Kuzvarira	a practice where in-laws marry their daughters to wealthy men
Mabwidi	people of Malawian origins
Maiguru	sister-in-law
Makandiwa	contour ridges
Makorokoza	gold panners
Makuvi	Wetlands
Mhashu	Locusts
Mifuku	holes in the sand of dry streams or river beds
Moda kutikovedza	you want to kill us
Mombe mbiri nemadhongi mashanu	two oxen and five donkeys
Mubwidi	a person of Malawian origins
Mubwidi akazviita	a person of Malawian origins won awards
Muenzi	Visitor
Mufushwa	dried vegetables
Mukoma	Brother
Mupfuuri hapedzi dura	assist the less fortunate
Murimi	Farmer
Musheche	sandy soil
Mwanangu	my son/daughter
Mzansi	South Africa
Pamuchawa	home of a person of Malawian origins
Pasichigare	long ago
Rukuvhute	umbilical cord
Ruware	rock outcrop
Sango rinopa waneta	the forest rewards those who persevere

Shaya uripo ngwena yepazambuko	persevere like a crocodile at a crossing point
Shona	largest ethnic group in Zimbabwe
Simbe	lazy farmer
Simbe hatidi pamusha pano	we don't tolerate laziness at this home
Shumba	Lion
Tingadzidza chinyi kumubwidi	what can we learn from a person of Malawian origins
VaMamvura	Waterman
Varungu	Whites
Vene vevhu	owners of the soil
Uhurudza	productive farming
Zvemonomono siyanayi nazvo	refrain from practising monoculture
Zvikomo zvishava	brown hills
Zvinoda akazviona	if only you witnessed it
Zvitokoloshi	Goblins